A Quest for Quality Wine, Every Time

Joyce Steakley • Bruce Steakley

A Quest for Quality Wine, Every Time

A Guide to Root Cause Analysis

 Springer

Joyce Steakley
Retired, Lockheed Martin
Cupertino, CA, USA

Bruce Steakley
Retired, Lockheed Martin
Cupertino, CA, USA

ISBN 978-3-030-34002-5   ISBN 978-3-030-34000-1   (eBook)
https://doi.org/10.1007/978-3-030-34000-1

This Springer imprint is published by the registered company Springer Nature Switzerland AG
The registered company address is: Gewerbestrasse 11, 6330 Cham, Switzerland

To my husband and co-author Bruce, who was the driving force throughout this effort and who did most of the writing of this book. He worked nearly every day for over 2 years and was never discouraged or distracted. His character, passion, and work ethic are traits that I truly admire.

To my wife and co-author Joyce, whose experience and knowledge have expanded my horizons and enabled this book. She has been the lead in our vineyard management and chief winemaker in our own winery operations. Her willingness to share her expertise and agility in a materials and process laboratory has elevated the quality of our own winemaking efforts to the highest levels.

To our wonderful daughters, Kathryn and Danielle. Their diverse accomplishments inspire us daily and fill our lives with joy.

Acknowledgements

Writing and publication would not possible without inspiration and assistance from many people. Much of our grape growing and winemaking knowledge has been learned through the exceptional online and on-site lab classes offered by the U.C. Davis Viticulture and Enology Department extension program. We are grateful to Howard Sawhill for his enthusiastic discussion and allowance to use his pet name the "BowWow" method for the process of searching and comparing the best of the best and worst of the worst attributes for effective problem-solving. We would like to acknowledge insightful comments from Richard Vassar on what makes compelling ideas and convincing material. We are also thankful for Jim Hawley's perceptive comments on winemaking, problem-solving, and excellent reference material. We would like to extend our appreciation to Michael Martella and Michael Michaud, two outstanding commercial winemakers, for their comments and inspirational advice on winemaking and best practices. Thanks are also due to Eric Escola, an exceptional winemaker and engineer, for his stimulating winemaking conversations and proposal comments. We would also like to recognize Bruce Firestone for his willingness to review proposal ideas and Ann Cutner-Firestone for her pursuit of sustainable living approaches. Credit also goes to Rick and Marie Golobic for Joyce's opportunity to work in their winemaking equipment shop and learn from their knowledge of winemaking, particularly for home winemakers. An enthusiastic shout-out goes to our good friends and "cellar rats" Chris Kolberg and Greg Buchanan for their unwavering hands-on support throughout all phases of our grape growing and winemaking process. Thank you also goes to Jim Donahue for his insightful comments on book writing and publishing approaches. We would also like to thank Myoung Kang and Ian Oglesby for a few well-placed motivational and inspirational tips. Lastly, we are grateful to all our wonderful friends and family who come out to help us in the vineyard and winery every year.

About the Authors

Joyce is a retired Aerospace Executive with 30 years of experience in materials and processes, contamination control, operations, and engineering. She has several published papers in the field of spacecraft contamination control. After retirement, she worked in the wine industry in a tasting room, a winery lab, a home winemaking supply shop, and installed a couple of home vineyards.

Bruce is a retired Technical Fellow with over 35 years in aerospace developing ground, air, and space optical sensors and systems. He has been a research scientist, engineer, systems engineer, and program manager. He has published technical papers and taught optical systems engineering courses.

Joyce and Bruce both have Winemaking Certificates from the U.C. Davis School of Enology and Viticulture. Both of us have spent decades making things and solving problems. We currently plant, grow, maintain, and harvest grapes from our own vineyard. We crush, ferment, age, test, and bottle the wine in our own winery. We have done many things right and many things wrong. Some of our wines have won awards and others have not. We have learned from both. We are two winemakers on a Quest for Quality wine, Every Time.

Contents

Chapter 1
Quest for Quality Wine, Every Time.
Guide to Root Cause Analysis.
Introduction

Winemakers share the desire to make great wine. Figure 1.1 is a symbolic signpost that there are many different grape growers and winemakers all around the world. Quality will mean something different to each of them. And all of them, at some time or another, will face a quality problem. A mystery that gets in their way. So, when something goes wrong, it is important to determine what the problem is, correct it, and prevent it from happening again. Rather than try something and wait a year or two for results, and then find out it happened yet again, it is important to get it right. A fierce and unrelenting pursuit is required, but it still may not be enough.

What can we do? There surely must be clues as to what went wrong. Which clues are worth considering? Do you have the extra time to pursue multiple paths? Given millions of process variations, this can be a never-ending effort. There has been a long wonderful history of winemaking and the internet allows anyone to search for answers to almost everything. This is not a problem of finding facts and data. Nor is this a time for trial and error experimentation. This is a need for organization, interpretation, and good judgement. This guide will do just that. In Aerospace we learned the methodologies of root cause analysis to consider, evaluate, and determine all relevant possible causes to find the root cause and act. This can be applied to winemaking.

This winemaker's guide to root cause analysis (RCA) describes a systematic process to find the root cause and enable an orderly path to resolution. Even though each winemaker is a unique individual and creative artist, there are basic fundamentals that we will share in this book. These help us all avoid flaws, solve problems, and make better and better wine. Whether you are a beginner or experienced winemaker, the logic in this process can help you. We will discuss a systematic method that is thorough and yet simple to perform. This process will distill daunting complex situations into clear positive results.

Before we start down the road of improving the quality of your wine, we begin our journey with a positive indicator of the quality improvement in our own wine. We believe that one of the most important indicators of the quality of your own wine

J. Steakley, B. Steakley, *A Quest for Quality Wine, Every Time*,
https://doi.org/10.1007/978-3-030-34000-1_1

Fig. 1.1 Whether you are new or old, commercial or amateur, big or small, you want to make quality wine

is your own assessment. Do you yourself like it? In addition, evaluations by expert wine judges is another good indicator. Figure 1.2 is a plot of critical scores given to five vintages of our Cabernet Sauvignon by various wine judges. We like to think that the quality of our Cabernet is improving. The scores are from blind tastings, independent, and probably less biased than our own. Critical scores and comments about the look, smell, and taste of your wine provide feedback on how you are doing and what path to take in the vineyard and winery to improve next season's vintage. We like continuous learning and use rational and systematic methods. We look for quality improvements around every corner. At the same time, we strive to include family and friends and bring some fun into the process. Now let us explain a set of powerful tools and embark on our quest.

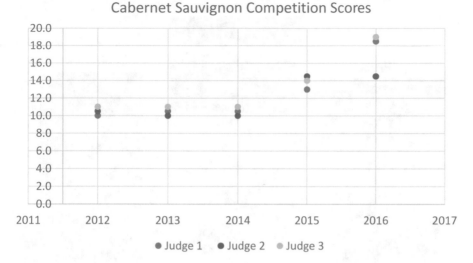

Fig. 1.2 Independent Wine Judges scores provide one type of objective evidence of the quality of your wine. Patterns will emerge, and trends will be revealed. Use rational systematic methods and solicit feedback along the way

These tools can be applied to any winemaking problem you might encounter, red or white. These techniques have wide applicability to almost any discipline. We will include select examples to illustrate a wide variety of applications. Our intention is not to provide an example of every detailed situation, but a sufficient number with adequate variety to clearly explain the process.

The universe of winemaking is broken down into six areas of interest as shown in Fig. 1.3. The first is process. These are the methods and procedures used to create the wine and finished product ready to drink or store for later. The second covers the various materials used to make the wine and come in contact with the wine throughout the process. The third includes all of the equipment that aids and enables the winemaking from vines to bottle. The fourth is the people. Mother Nature supplies the grapes, but people conduct the harvest and make the wine. The fifth area includes the measurements made by the winemaker to analyze conditions and drive decisions. The sixth and last major area is the environment of conditions that begins with the vineyard soil, climate, winemaking chemistry and finally storage.

The following chapters will describe this process in detail and provide examples that are end to end scenarios for various red and white wine flaws. We will discuss how to identify and detect characteristics. There will be a description of how to evaluate and analyze these observables followed by options to repair the wine. Finally, we include ideas on how to prevent or mitigate possibilities of these problems from happening again.

Apply this method and salvage a wine or fix a process for next time. Use these techniques and in the long run you will save time and resources. These investigation approaches can solve any complex problem in any discipline. Follow this guide and you will improve your chances of making quality wine, every time!

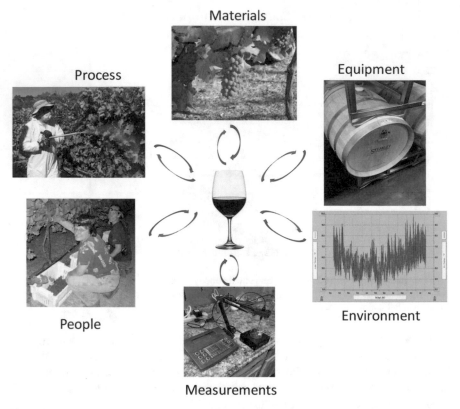

Fig. 1.3 Our quest for quality wine looks at the world of winemaking through a root cause analysis divided into six categories

Chapter 2
Quality Wine: Degree of Excellence and Distinctive Characteristics

2.1 What Is Quality?

What does quality mean? The word is used all the time and it is found almost everywhere. It is discussed by many and is still so vague. Quality wine is a worthy goal, but what constitutes quality can be hard to grasp. It can mean something different to two different people. A certain wine might be called high quality by one but ridiculed as low quality by another! Many strive for gold medals year after year because it is an endorsement of quality from independent judges. Are gold medals what we want? Quality wine is what we want. Could quality be such a nebulous word that we will never really settle on what it means? We do not think so. Let us first figure out what quality wine means to us. We will explain what it means, teach you tools to figure it out for yourself, and provide examples of what wine experts think.

Quality has a couple of commonly used definitions. One definition is that the quality of something is its special character and essential features. This might be the color, aroma, or taste of a wine. Another definition is that the quality of something is its degree of excellence or level of superiority. This means the intensity or severity of wine characteristics such as color, aroma, or taste. There might be many different characteristics of your own quality definition and there may be many levels to the degree or severity of these characters you want in your quality definition. The distinction of the two definitions is important to understand. We recommend incorporating elements of both in a winemaker's quality definition. Having both aspects help you drive more efficiently and effectively toward solving problems you encounter. It helps you understand if you have achieved the broader goal or not. It provides signposts to guide you on your own quest for making quality wine.

Quality can have many layers. The word can be as complex as the wonderful, complex wines we enjoy. Coming up with a well thought out definition is very helpful. You may want it to look good, smell nice, and taste good. You also may want to win lots of awards, fetch a high price, and sell many cases. Think carefully about

J. Steakley, B. Steakley, *A Quest for Quality Wine, Every Time*,
https://doi.org/10.1007/978-3-030-34000-1_2

Fig. 2.1 Do you have the right quality goal? For our 2012 wine we got a gold for the label and bronze for the wine inside

what quality means to you. This is the first step in figuring out your own quality goals. This will help in solving winemaking problems more efficiently and effectively.

One year, we happened to get a gold ribbon for our label and bronze for the wine. Figure 2.1 shows our award certificate. My goodness! We smiled, then sighed, then smiled again. We were hoping the wine was better. We were happy and disappointed at the same time. We were not focused on the wine. We did not have well-understood goals while making the wine. The goal of this book is to teach systematic, rational methods to solve wine quality problems. We believe making quality wine begins with establishing a well-understood quality goal.

You may want your quality wines to be an expression of the terroir where the grapes are grown. You also may want to retain characteristics of the wine your loyal customers have come to expect. These quality goals may appear different but could simultaneously be different layers of the one wine. We need a definition of quality that is sufficiently clear and deep that we know where we are headed. We also need a definition that we can hold up to scrutiny and know what is missing when we have not achieved it. Not only are we going to teach you some wine problem-solving methods, we are going to show you how to make quality wine by your own definition.

2.2   Get to Personally Know the Quality of Your Own Wine

First develop your own sense of the quality of your own wine. Make your wine with the intention to improve its quality. This requires a good quality definition, understanding the quality of the wine you are making, and your quality goals for the future. The character qualities of your wine and its level of superiority can both be goals. Keep each in mind as we work our way through this section. Keep both in mind as you begin your quest to improve the quality of your wine. We want to achieve both aspects of the broader quality definition. We want you to develop details of both aspects of your own quality definition. Let us discuss ways to develop your own opinions on whether your wine satisfies your own quality goals. Pour a glass of your wine and let us take a good look.

Look at the color of your wine and get to know its many color characteristics. Look at your wine. Look at it a lot. Do you like the color? Is it changing year to year? Get to know the look of your wine. We strongly advocate viewing in controlled lighting conditions. Tilt the glass over a white background as shown in Fig. 2.2. Note how the color at the edges is lighter as the thickness of the wine gets thinner. Note the deep colors closer to the center. Make sure the room and the wine are well lit. This will help you see the true color of the wine and allow you to repeat this color inspection throughout the winemaking steps and from year to year. You can take pictures and note the time and date. Use a color wheel or other reference like the Wine Folly poster to record the colors you observe. This has become very convenient with a camera on almost every phone. You can track changes throughout the various winemaking steps, through aging, and from vintage to vintage. Note the subtle hues and shades. Take mental and written notes from time to time and begin a history of the color of your wine. Try to correlate color with aroma and taste to have a better visual cue of the quality, particularly as the wine ages.

Now take a few minutes to smell the aroma of your wine. Here again, we recommend you smell and sense the aroma frequently. This too should be in a controlled and convenient setting such as in the winery as shown in Fig. 2.3. Develop a memory and history of the smell of your wine. Is it pleasing? Does it make you want to taste it? Get to know the smell of your wine. Try to record as many descriptors as possible so that over time you will be able to pinpoint the characteristics that contribute most to your favorite, high quality, wine.

There is a wonderful aroma wheel[1] that was developed by Anne C. Noble that is a great tool to help enhance your own capabilities to remember and talk about the aromas of your wine. You will be able to note how it changes or evolves throughout

Fig. 2.2 View and note your wine color and hue in controlled conditions

---

[1] Ann Noble's wine aroma wheel is available from the U.C. Davis Bookstore.

Fig. 2.3 Smell the aroma
of your wine. Smell it a lot.
Get to know the aromas of
your wine. Do you like it?

the winemaking steps and from year to year. You will be able to compare your aromas with other aromas of wines you like. This will allow you to "dial in" the characteristics you value most, like the presence or absence of aromas such as vanilla (using oak barrels) or bell pepper (with vine canopy management including leaf pulling).

Let us take a taste. It is important (and fun) to taste your wine frequently—from the early grape juice phase to the wine bottling step. Figure 2.4 shows tasting in the winery that can be conveniently associated with other periodic events such as topping off, racking, and or testing of wines. Save a few bottles in the cellar to taste prior vintages and note how they evolve. Develop a memory and history of your wine tastes. Describe the complex tastes in terms of sweet, sour, salt, bitter, and umami (savory). How is the acid or residual sugar? Note the body of the wine. Note the mouthfeel and the finish. Mouthfeel includes perceptions of astringency, dryness, viscosity, burning, heat, coolness, body, and prickling. Note the taste of your first sip and how it evolves in the glass over time.

It will take time and practice to understand your own wine. Understand and enjoy the wine you make. This is where we like to start in establishing your own wine quality definition and goals. There are a number of tools and techniques to help with this. As part of this chapter, we are now going to cover more formal numerical wine evaluation systems, expert wine critic feedback, wine labels, and wine quality proxies.

Fig. 2.4 Taste your wine. Taste it a lot. Note how it changes throughout the winemaking process. Get to know the taste of your wine

2.3   Quality Assessments and Indicators

We have broken out four types of aids to build your own definition of quality. Table 2.1 lists the different types which span a broad range of sources. The first is a system of numerical evaluation of wine characteristics created by Dr. Maynard A. Amerine. The second is a single numerical scoring example from some of the world's finest expert wine critics. The third comes from national level wine label and production laws and regulations. The fourth is a look at wine quality proxies and how they correlate with one another. Together, they will provide you with many resources and examples to create a broad and deep understanding of wine quality. This in turn is a sturdy launching point for setting your own quality goals, making your own quality wine, and solving any wine problems that get in your way. These all are resources you can use to evaluate your wine and improve the quality each year. They are ways to provide objective feedback as to whether you have achieved your quality goals. They may be wine consumer oriented, but numerical scoring of wine characteristics and comparisons with other known quality wine standards are also helpful to the winemaker. They will not only help develop your own quality definition, they will enable a breakout and refinement of quality attributes and layers. There are many different quality attributes or proxies that make up a quality wine. Whether you learn from all of them, some of them, or just a few; the extent to which you establish your quality goals early will improve the quality of your wine later. The following table provides descriptions, benefits, risks, and weaknesses of

Table 2.1 Quality indicators can be found in many forms

Type	Description	Benefit	Risks and Weakness
1. Numerical scores for various wine characteristics	• A system that assigns numerical scores for wine attributers such as the Dr. Amerine's system (Amerine, Roessler, & Fillipello, 1959) • Quantitative assessments for wine characteristics such as color, aroma, acid, sugar, taste, finish • Typically contains associated description of score basis	• This provides insight into different sensory and component aspects of the wine • Can guide wine consumer before tasting for themselves • Can help winemaker by virtue of evaluating isolated wine characteristics that may themselves be driven by particular grape or vinification methods	• Assessment of a few characteristics, but still a judgement by individual or team of people with varying levels of experience and capabilities • Judges will have their own palates or preferences and it may not fit yours
2. Single numerical scores and commentary from wine experts	• Methods that assign a single numerical quality score to the wine • Each may use their own respective min and max range and basis for the scoring assignments or bins	• Quantitative fast assessment for consumers • Easy to track, make comparisons, advertise, and brag about • Supporting comments provide better understanding of basis	• No insight on how to fix low-scoring wines nor maintain quality of high-scoring wines • Wine quality hard to reduce to single number • Judge or judges may not share same palate as you
3. National wine label and production laws and regulations	• Many countries have national level wine rules that govern labeling and production of wine. Most have geographic based designations. Some have additional elements addressing viticulture practice and vinification	• Worthy goals of providing consumers with trusted quality indicators on labels • Many offer quality level designations on the label enabling some wine assessment before you buy • Provides uniform country or regionwide rules	• Rules can be confusing. They can be opaque or obscure to the average consumer • Rules can and will vary from country to country and region to region • Some winemakers may not follow the labeling rules (knowingly or unknowingly).
4. Wine quality proxies correlation study (Hopfer, Nelson, Ebeler, & Heymann, 2015)	• Correlated quality proxies for Cabernet Sauvignon with sensory attributes and chemical composition	• Scientific study that shows correlation between wine expert scores, retail prices, vintage, wine region, sensory attributes, volatile compounds, and chemical composition	• This is an exceptional correlation analysis and useful database; however, it does not address (nor make any claims to address) quality problem solving cause and effect

These range from characteristic numerical evaluation, wine critic reviews, wine law regulations, and correlations amongst characteristics and quality indicators two numerical scoring approaches, quality elements found on wine bottle labels, and a magnificent study on the degree of correlation between quality attributes. This last item connects some of the dots between various quality proxies such as sensory characteristics, (Amerine & Winkler, 1944) wine chemistry, vintage, region, and price for a group of Cabernet Sauvignon wines. These four items provide a broad and thorough view of wine quality.

2.3.1 Maynard Amerine and the University of California at Davis 20-Point Scale

Dr. Maynard Amerine was an exceptional Professor of Enology at the University of California at Davis. Dr. Amerine and his team created one of the earliest and most notable 20-Point Scale Wine evaluation system (Amerine & Winkler, 1944). This historical paper on wine evaluation was a masterpiece in explaining how to evaluate wines. This was the genesis of what is also referred to as the University of California at Davis 20-Point Scale System Organoleptic Evaluation Scoring Guide for Wine. It was a method for rating the large number of experimental wines that were being produced at the university. Wine was and is meant to be enjoyed and shared with friends. But to the winemaker, tools are needed that can deepen understanding, enable repeatability, and provide accurate unbiased assessments. This is a very tall order. It is also hard to believe that such tools exist for a beautiful enjoyable thing such as wine. We believe they do exist and Dr. Amerine's 20-point scale is shown in summary cumulative form in Table 2.2.

This approach evaluates the wine and assigns points in each of ten subcategories. They are added up to make a maximum possible score of 20 points. The integrated score for a given wine is designed such that the total score for the wine indicates the integrated quality of the wine. The genius of Dr. Amerine's system is that it builds upon the lower level scores within each of the ten subcategories. Each subcategory grade or value relates to discernable separate characteristics of the wine. Even the

Table 2.2 Quality levels inform the winemaker

Score	Quality level of wine characteristics
17–20	Wines of outstanding characteristics having no defects
13–16	Standard wines with neither outstanding character or defect
9–12	Wines of commercial acceptability with noticeable defects
5–8	Wines below commercial acceptability
1–5	Completely spoiled wines

They are intended to be applied without bias and consistently from wine to wine and vintage to vintage

Table 2.3 Scoring rules provide objectivity and repeatability to the process

Category and Score	Score assignment rules
Appearance (0–2)	0 if cloudy, 1 if it is essentially clear, and not flashing or light reflections, 2 if it is brilliant with no dullness, no murkiness, and no particles nor sediment
Color (0–2)	Acceptability or goodness of color depends on the grape varietal and on the age of the wine. Whites may be yellow, gold, or straw, Pinot Noir may be light, Cabernet may be deep red. Brown tints or deep red may be faults
Aroma and bouquet (0–4)	Aroma is the sense of smell originating from grape variety characteristics. The intensity maybe light (2), medium (3) or high (4) with flaws of being too alcoholic, woody, moldy, or corked getting a 0 or 1
Volatile acidity (0–2)	If the wine has no vinegar smell, it gets a 2. If it has a slight vinegar smell, it gets a 1. If it has a strong vinegar smell, it gets a 0
Total acidity (0–2)	This is mostly felt in the mouth around the edges of the tongue. If it is low, the wine is flat or flabby. If it is very sharp and highly zingy, this could get a 2
Sweetness/ sugar (0–1)	Sugar and total acidity are evaluated together and should be balanced. If it is overly sweet or overly dry for its style, then it should get a 0
Body (0–1)	This is the viscosity of the wine and is referred to as mouthfeel. Too thin like water, should get a 0
Flavor (0–1)	The flavor should correspond with the bouquet and aroma. If it is clean and fruity, and full or balanced it should get a 1. If it has any metallic, steamy, or non-varietal character, it should get a 0
Astringency (0–2)	This category should also be judged in accordance with the age of the wine and its particular variety. If the wine has ideal astringency, it is mellow, soft, velvety, and has a roundness and it should get a 2. A young wine is not downgraded for natural tannins
General quality (0–2)	This one category is subjective and allows for the judge to adjust the score on the basis of the wine's total performance and impression

Stratification with scoring levels provide further quality insight to the winemaker

variation of personal taste can be overcome with training, repetition, and comparison against standards.

The unique subcategories are shown in Table 2.3. Note how the categories allow greater objectivity to the assessment. The characteristics and scoring will depend upon the grape varietal, age, and style of wine. The resolution of the integrated score is good by virtue of having ten subcategories and each with its own valuation rules. Reliability and confidence in the score are achieved by training the judges with controlled evaluation conditions, repetition against known standards, and using multiple judges and multiple blind tastings.

You as the winemaker can use the original subcategories and scoring levels or adjust to fit your own preferences. We have tailored this approach ever so slightly to better fit what we think fits our ability to discern variations. The key is to breakout the evaluation into discrete characteristics and evaluate each. We also work to evaluate the wine characteristics separately and independently from each other. Sometimes Bruce's assessments (shown in Fig. 2.5) and Joyce's assessments (shown in Fig. 2.6) are similar and sometimes they are very different. This leads to a stronger and better

Wine:		Cab Sau. 2010		Merlot 2010		50/50	
		Score	Comments	Score	Comments	Score	Comments
Color	(0-1)	1	Dark Ruby Garnet	1	Red.	1	Dark Ruby Red Garnet
Clarity	(0-1)	1	Clarity good	1	Clear.	1	Clarity good
Aroma/Bouquet	(0-5)	3	Vegetative Licorish Vanilla	4	Berry strong Vanilla	4	Berry Vanilla like Veg, Tobacco
Taste Balance	(0-5)	3-4	Berry	4	Vanilla	5	Mild, Smith
Sugar	(0-1)	1	dry	1	Dry	1	Dy
Acidity	(0-2)	2		2		2	
Body	(0-2)	2		2		2	
Tannin	(0-1)	1	pucker	1		1	
General Quality	(0-2)	1		1		1	
Totals	20	15-16		17		18	

60 gal = 300 bottles = 25 cases

Fig. 2.5 We apply the Amerine/UC Davis 20-point system to our own wine evaluation. We have created our own tailored version based upon the general set of original categories. Bruce's notes on these three different wines are a bit cryptic and brief. You can do better

assessment.[2] We keep records and track the trends. Establish your goals early and keep checking to see if you have met them. Over time and with repeated evaluations, the winemaker can correlate winemaking methods and characteristics to the quality of the finished wine. In the beginning this helps understand what quality means and to compare with other wines and the quality of their respective subcategories.

Whether you are a commercial winemaker or home winemaker, there are various independent groups available to evaluate your wines and provide feedback. Beautiful blue ribbons and gold medals can be fun. They make you feel good and you can show off a little. However, as a winemaker, blue and gold ribbons may not give you all you need, and they may not show up year after year. We do want critical feedback. We need independent feedback and critical discerning comments. We can grow from understanding what parts of our wine are bad. We can learn from which aspects are of poor quality. Numerical scores help us understand how they rank against other wines and allow us to keep quantitative historical records. We can look

---

[2] Note that a few minor arguments and accusations may ensue but work to get past these!

Evaluator: __Joyce__   Date: __4-1-2012__

Wine:		2010 Cab Sauv		2010 Merlot		50/50 Blend	
		Score	Comments	Score	Comments	Score	Comments
Color	(0-1)	1	very dark	1	lighter	1	dark purple
Clarity	(0-1)	1	clear	1	clear	1	clear
Aroma/Bouquet	(0-5)	3	tobacco veggie licorice chocolate	4	fruity berries	4	tobacco berry veggie
Taste Balance	(0-5)	3		4	vanilla smooth	4	vanilla
Sugar	(0-1)	1		1		1	
Acidity	(0-2)	2	pucker	2	pucker	2	pucker
Body	(0-2)	2	nice mouth feel	2	nice mouth feel	2	nice
Tannin	(0-1)	1		1		1	
General Quality	(0-2)	1		1		1	
Totals	20	15		17		17	

Fig. 2.6 Evaluate your wines multiple times with multiple people. Joyce and Bruce do not always agree. That is ok!

for trends and patterns. Comments against subcategories provide a deeper level of evaluation. Commenting and scoring on subcategories like color, aroma, and taste can even provide the winemaker with more precise direction on where and how to improve. Now if we use the same winemaking recipe year after year, why do we not get the same result? Oh, my goodness, there are so many variables that may have changed; the same recipe may need tuning, adjustment, tailoring, and enhancements just to maintain the same degree of quality.

Let us look to wine experts for useful guidance and ideas in developing your own quality definition and goals. Many wine retailers, distributers, and producers look forward to getting high scores and positive comments from Robert Parker, Jancis Robinson, the Wine Spectator, or some other respected wine critic. Consider each of these as a possible source for inspiration in defining your own high-quality goals. Tasting other wines with high-quality scores and comparing with your own can be a useful path to defining your own quality goals.

The methods and tools explained in this book are primarily focused on wine production, not wine consumption. However, we can also take cues from experts in the wine consumer side. Personal taste in wine can be a very individual thing. It can be hugely different over the same wine. Because of this variation, we look to multiple sources and independence. This improves our chances of getting unbiased objective positions. Wine critics are often targeting wine consumers; however, the winemaker can still benefit from the findings of expert wine critics. We have a lot to learn from expert wine critics. Insight into establishing your own wine quality goals may be found here. The high scores themselves will not explicitly guide your winemaking process, but they are feedback as to whether we have solved your own problems and achieved your goal of making a quality wine.

We not only recommend making a wine that you personally enjoy, we also recommend getting independent feedback on the quality of your wine from expert wine critics. We enjoy serving our family wine to our family and friends. They say they like it; however, we are never quite sure if they really like it or if they are just being nice. They are our friends! We certainly appreciate and enjoy getting positive feedback from friends and family, but as winemakers, we also need critical and independent feedback. We have discussed an exceptional and original piece of work done by one of the greatest wine enologists, Dr. Maynard Amerine. Let us now turn our attention toward a few exceptional wine critics and their respective scoring and wine evaluation methods. One of the common features of this group is that they provide a single numerical score for the quality of the wine and some commentary and discussion of wine characteristics. These can be another source of establishing and refining your own wine quality definition.

2.3.2 Quality of a Wine Determined by Experts and a Single Numerical Score

Wine critics and popular wine evaluation systems directly help the consumer, but do not forget that they can also help the winemaker. The various attributes of wines, their respective grape growing conditions, the years they were made, and how these compare with your own wine situation, all will help you associate things that can be relevant to your own wine. Understanding how wine critics rate the quality of other wines and how these compare to yours, is a thread for sewing together your own wine quality.

2.3.2.1 Robert Parker's Numerical Scoring System

One special example is that of one of the most influential wine critics, Robert Parker (Parker & Rovani, 2002). His system was first published in the early 1970s and provided a simple single numerical score for the consumer to easily assess the quality of a particular wine. We applaud the quantitative aspect because it provides so many benefits. It provides the ability to quickly rank the quality of wine. It makes it easy to advertise a high score quality. If you are a seller, you might ask for a higher price. If you are a buyer you might expect a higher quality but unfortunately, you will probably pay a higher price. There are so many wine producers from around the world and so many different styles of wine. It can be overwhelming to try and figure out which wine to choose. It makes it easy to track wine quality from year to year and establish trends. Hooray for a step toward quantifying quality. This is primarily wine consumer oriented; however, there are also secondary benefits to winemakers.

Robert Parker himself is the principal judge of these wine scores, and it is a matter of his personal tastes. However, he does provide an excellent basis for how he grades. He describes what he looks for in the various quality levels in wines. These are the ability to please both the palate and the intellect, the ability to hold the taster's interest, the ability to hold intense aromas and flavors without heaviness, the ability of wine to taste better with each sip, the ability of a wine to improve with age, and the ability to offer a singular personality. These are so very important and illustrate the many different facets. The score makes it quantifiable, but the challenge is still that they can be subjective and contain the biases of one individual.

The enjoyment of a wine can come from many different sensory characteristics, as well as the terroir of the grape region, and history of the wine producer. Not only the sensory aspects of the wine itself, there is the consumers own awareness and appreciation for where the wine came from, how it was made, and the long history behind it. The enjoyment of a high-quality glass of wine comes from all of these. As you begin your own quest for making quality wine, these kinds of quality characteristics are excellent starting points. These are worthy quality goals. We have stated a few times that they are consumer oriented; however, understanding consumer quality preferences helps the winemaker as well. A happy and satisfied wine consumer is good place to be.

Robert Parker's scoring system ranges from 50 to 100 and is described in Table 2.4. This is wine buyer oriented, but it still provides some insight to the winemaker. It can be a source of attributes a winemaker may choose to achieve in their own wine. Note how it stratifies the quality a wine. It enables comparison with other wines. As a quality winemaker, one goal could be to achieve high scores. This should not necessarily be the only goal, but it could very well be one aspect of your quality goals. The winemaker will also be looking for feedback on how to improve

Table 2.4 Robert Parker's numerical scoring has values that range from 50 to 100 (Parker & Rovani, 2002)

Score	Description for Robert Parker's scoring range
96–100	Extraordinary
90–95	Outstanding
80–89	Barely above average to very good
70–79	Average
60–69	Below average
50–59	Unacceptable

These are six bins of quality spanning well below to well above average and maintain the quality, but this is reasonable place to go to develop your own quality definitions.

Note how the numerical score range is from 50 to 100. With expert senses, lots of samples, and years of tasting, it is quite possible that a talent like Robert Parker might be able to resolve 51 different bins over that numerical range. However, for most of us, or the typical winemaker, this quantitative sensory resolution is not likely. This might improve if one spent a lot of time getting calibrated to Robert Parker's evaluations by personally tasting and evaluating the same wines and comparing the scores and assessments. This would be very impractical and time-consuming. However, it is still helpful to build a reasonable level of personal calibration to a trusted wine critic's evaluation by sampling a few wines. A winemaker can gather useful insight from Robert Parker's scoring of similar style wines and associated commentary. This is a necessary beginning or aspect of the process, but it is not sufficient nor really provides any details to the understanding of the root cause of your own wine quality shortfalls.

It is important for you to understand what you want to achieve exactly for your quality wine. The quality wine aspects that Robert Parker might be precisely what you are looking for nor what you are trying to achieve. Let us take a quick look at another numerical scoring system from another notable wine critic and explore the similarities and differences.

2.3.2.2 Jancis Robinson's 20-Point Numerical Scoring System

Hugh Johnson, Jancis Robinson, and Mitchell Beazley are superstars in the wine world. They co-authored a great manifesto on wine, The World Atlas of Wine (Johnson, Robinson, & Beazley, 2007). Jancis Robinson has written about understanding wine, tasting wine, and enjoying wine. She is a recognized expert on wine all around the world and has received numerous prestigious awards. She has an informative website JancisRobinson.com (Robinson, 2019) that includes daily updates on what is happening in wine. The website includes a discussion of how she and her group score and evaluate wines. She mentions that one of her tasters has a similar palate as she does. Tuck this away in your winemaking kit because there will come a time when you too may taste and collaborate with others on making, tasting, and enjoying wine. It is good to know and get calibrated to the palate of other experts.

The calibration concept is used in science, engineering, and making complex things. It is determining the relationship of something to something else that has a well-understand value. This may involve figuring out adjustments or corrections to offset a value to bring it closer to the truth. If you follow a notable wine critic and enjoy their highly rated wines, you are beginning to get calibrated to their tastes. If you dislike the wines that the wine critic gives low scores, you are also adding confidence. If the wine critic pans another wine, you may have reasonable assurance that you will not like it either. You are getting "calibrated" to that particular wine critic's scores. We will discuss more about calibration in upcoming chapters on the use of measurements for providing scientific feedback on the characteristics of your wine. In your quest for making quality wine, it is not only good to start with clearly defined quality goals, it is also good to get independent feedback from other wine experts. This can be directly from their tasting your wine or indirectly from inferring evaluations of similar style wines. It is helpful to understand what they mean by quality and how close your wine fits your quality goals.

Jancis Robinson is a long-time world recognized wine critic and trusted source. Her numerical scores and descriptions are summarized in Table 2.5. Would not it be fun to make a "Humdinger" or extremely disappointing to make something that was "Deadly dull"? The colorful phrases do indicate what is above average, average, and below. Note how the winemaker can receive qualitative critical feedback that may be fun and informative at the same time. A winemaker can be happy having made a quality wine if it receives a "Truly Exceptional" evaluation. If your wine is borderline faulty or unbalanced, this is bad and unacceptable. The brief description in itself has some value; however, the winemaker can glean even more insight by personally tasting and evaluating the same wines as the wine critic and comparing scores. As we discussed above, you can calibrate yourself to Jancis Robinson's taste.

Evaluating a wine with great repeatability and precision by viewing, smelling, and tasting may not work well for everyone. However, for a trained and talented wine critic, this can be effective. It is probably a stretch of credibility to argue that most people can discern 51 discrete bins, let alone 20. Expert wine critics can do this for you and enhance the information and tools to develop your own wine quality goals. This can build upon your own senses to appreciate and discriminate your own wine. The long but rewarding process needs you to set your own quality goals, use your own senses to identify these, and use grape and wine chemistry to further increase your objective understanding of the quality of your wine. Would not it be nice to have a wine quality indicator you can see before you buy and open the bottle? Of course, there is. It would also be nice if these attributes could help me as a

Table 2.5 Jancis Robinson provides another example of a numerical scoring system and associated descriptions (Robinson, 2019)

Score	Description of score
20	Truly exceptional
19	A humdinger
18	A cut above superior
17	Superior
16	Distinguished
15	Average, a perfectly nice drink with no faults but not much excitement
14	Deadly dull
13	Borderline faulty or no balance
12	Faulty or unbalanced

Her descriptions are fun and informative at the same time! They provide critical review in a playful way

winemaker to establish quality goals and refine my winemaking process to make higher quality wine. Let us discuss wine laws, labeling, and then move onto how these items might correlate with quality.

2.3.3   Quality Found in Wine Laws and Labeling

The world of wine has an inspiring history of improving wine quality in many aspects. At the risk of entering into a mind-numbing and tedious narrative, we will just take a quick look at discernible quality features found in wine laws around the world. Think about a recent quality wine you have enjoyed, or better yet; grab a glass, get comfortable, and read on. The primary objective of wine production and labeling laws may be to help inform consumers, but there is also a quality spinoff. This is one that is relevant to the winemaker and wine producer. Wine laws are primarily intended to regulate production, sales, and labeling of wine. They are intended to combat fraud by documenting requirements when producing wine, selling wine, and designating certain attributes on labels. They are also a good source of wine quality attributes for both the consumer and the vintner. Knowing about the varieties within the wine, the regions where the grapes were grown, and the vintage when the grapes were harvested are all very important descriptors. They not only imply a lot about the basic taste, but they also infer something about the wine's quality. Information about the wine producer and its purchase price also enables further insight into the wine quality. Extrapolating quality implications from this information is far from perfect, but it gets you closer.

Wine laws and rules vary considerably around the world. They are different between countries, regions, states, and local communities. Figure 2.7 tabulates examples of wine grape growing and aging regulations. The French regulations are explicitly specified whereas the US regulations not. Understanding and complying with these wine production regulations and wine labeling laws is required if you

Fig. 2.7 Examples of significant wine law regulation differences that impact quality. The French system regulates certain aspects that the US system does not. Both produce high-quality wines

Wine Laws		
Phase	Regulate	Choice
Grow – variety by region	🇫🇷	🇺🇸
Grow -irrigation	🇫🇷	🇺🇸
Grow – Max yield per acre	🇫🇷	🇺🇸
Harvest	🇫🇷	🇺🇸
Age	🇫🇷	🇺🇸

wish to produce and commercially sell your wine. We will describe these rules and how they relate to quality. Understanding these requirements can be used for early development of your quality goals and improve your winemaking process. We are not going to cover all wine regions of the world, but we will take a quick look at the USA, France, and Australia. We will highlight similarities and differences between the old and new worlds of wine.

The grape growing region is a foremost indicator of quality potential in the wine regions around the world. This is a precisely defined geographical area where only wines produced from grapes grown within can claim this appellation and certain regulations and rules apply. When the soil, sun, and climate conditions are combined they drive wine character, and this is the essence of the French term terroir. The US TTB has a similar term for grape growing regions called American Viticulture Areas (AVAs). The Australian system designates wine grape growing regions with a Geographical Indication (GI) such as states, zones, regions, subregion, or vineyards also by distinct characteristics within a boundary. The grape growing region is arguably one of the biggest drivers in finished wine quality. This is recognized by many wine producing countries around the world.

Wine and quality characteristics are heavily influenced by grape variety. The French and European systems have restrictions on grape variety for given areas that are permitted for winemaking. The US and Australian systems allow the grape grower and winemaker to make a varietal choice. Knowing a grape growing region, its soils, its climate, and history of wines made from grapes grown within is a powerful indicator of likely wine quality. By virtue of the French system identifying grapes used in high-quality wines and then regulating the use of these same grapes may still allow production of quality wines, but it does limit choices for winemakers using grapes from these regions. Figure 2.8 tabulates select grape growing and winemaking phases intended to regulate high quality in the French system and allow choice in the US system.

Fig. 2.8 High quality maybe in the laws or it maybe achieved if you make the quality choice. Understanding the regulation differences leads to a better understanding of potential quality differences

Wine Laws		
Phase	High Quality	High Quality if make the quality choice
Grow – variety by region	🚩	🏴
Grow -irrigation	🚩	🏴
Grow – Max yield per acre	🚩	🏴
Harvest	🚩	🏴
Age	🚩	🏴

The yield maximum (or the maximum weight of grapes) per acre is set for red and white wines. Too high a yield may reduce the quality of the wine. For a given area, there is no US TTB-controlled limit on yield per acre. This is a choice and only limited by viticulture practice. There will be consequences on the quality of wine, but US producers are not required to divulge this information. A winemaker may be targeting higher volume, lower quality, and lower priced wines. Higher yields may result in lower quality if there are not compensating viticulture or winemaking practice which all will not be perfect and have limits.

Table 2.6 shows other areas where there are differences between the old world French system and new world US and Australian systems. The French old world system controls canopy management methods and timing, the design of the trellis system, and irrigation. There are no wine laws in US nor Australian systems controlling these aspects. Each of these can and will impact grape quality.

Table 2.6 Variations between old and new world wine production regulations illustrate evolving traditions, philosophy, and quality preferences (MacNeil, 2015)

Item	France (old world)[a]	USA (new world)	Australian (new world)	General quality
Grape growing region	The appellation d'Origine Contrôlée / Protégée (AOC/AOP) is the geographical area where wines produced from grapes grown within can claim the appellation	American Viticulture Area (AVA) is a wine grape growing region distinguished by geographic features, with boundaries defined by the Alcohol and Tobacco Tax and Trade Bureau (TTB) (Alcohol and Tobacco Tax and Trade Bureau, US Department of Treasury, 2019)	Wine Australia (Australian Wine, 2019)[b] designates wine grape growing regions with a Geographical Indication (GI) also by its distinct characteristics within a boundary	Old and new worlds all acknowledge that geography plays a major role in wine quality
Grape variety	For a given area, only a controlled list of grape varieties can be used and only in specified relative concentrations	There are no US TTB regulations on grape variety for a given area	There are no Wine Australia regulations on grape variety for a given area	Wine characteristics heavily influenced by grape variety
Yield per acre	The yield maximum (or the maximum weight of grapes) per acre is set for red and white wines	For a given area, there is no TTB-controlled limit on yield per acre	No Wine Australia government rules restricting yield per acre	Without offsetting methods, higher yield may lead to lower quality

(continued)

Table 2.6 (continued)

Item	France (old world)[a]	USA (new world)	Australian (new world)	General quality
Vineyard practices	Rules control canopy management, trellis design, irrigation, and more	For a given area, there is no TTB regulations on vineyard practice	No Wine Australia government rules restricting vineyard practices	Old world control, new world choice
Winemaking practices	Various vinification techniques such as adding sugar (chaptalization) or adding acids or aging periods are regulated	There are US TTB limitations on adding materials but none on aging	There are no Wine Australia government rules regarding wine vinification practice	Less regulations in new world systems
Tasting and chemical analysis	All AOC wines must be tasted to ensure they are appropriate for their type of wine and undergo chemical analysis to pass limits	No TTB representatives are required to taste the wine for certification or authentication. To conduct commercial winemaking operations the TTB has various requirements to be an approved bonded winery	There are no Wine Australia government regulations on taste testing. There may be audits with chemical analysis primarily for wine contamination issues	The French regulate tasting and testing their wines. The US and Australians leave more up to producers

[a]The appellation d'origine contrôlée (AOC); "protected designation of origin") is the French certification granted to certain French geographical indications for wines, cheeses, butters, and other agricultural products, all under the auspices of the government bureau Institut national des appellations d'origine, now called Institut national de l'origine et de la qualité (INAO)
[b]The Australian government authority on wine regulations is Wine Australia. https://www.wineaustralia.com/

The French system does contain regulations on vinification practices. The US TTB has restrictions on materials that can be added but no regulations on winemaking practice.

Labeling of wine has similarities and difference between the French, US, and Australian systems. Note in Table 2.7 how the old and new world systems take different approaches to address perceived quality and safety characteristics. There are no allergy warnings found in the French system; however, the US, and Australian labels must contain sulfide warnings. The Australians have also required warnings regarding the use of milk products. These are health oriented and not quality related. Both old and new world examples have very high percentage requirements which pay respects to the powerful impacts of the terroir of the regions to influence the ultimate quality of the finished wine. There are quality categories on French wine labels. Wine regions in France have their own respective quality classifications.

Table 2.7 Old and new world wine labeling laws approach quality and safety from different angles

Item	France (old world)	USA (new world)	Australian (new world)	General quality comment
Degree of alcohol on the label	All AOC areas have minimum alcohol thresholds, and some have maximum alcohol limits	All US made wine must show alcohol content and warning of the dangers of alcohol	Wine Australia requires alcohol content to be shown on the label	Similar between old and new world systems
Allergy warning on the label	No allergy warnings required by French AOC	The US TTB requires sulfite content warning on label (TTB Alcohol and Tobacco Tax and Trade Bureau, 2019)	Wine Australia requires wording indicating sulfide concentration and use of milk	Allergy warning required in new world systems
Grape variety on the label	No grape varieties required to be shown on label	≥75% of grapes used to make the wine must be of the declared variety and from identified appellation of origin	Optional, but if declared, must be ≥85%. If multiple varieties must list in descending order	Not in old but yes required in new world class of varietal wines
Vintage designation on the label	Hundred percent of the grapes must be harvested in the designated year	Ninety-five percent of the grapes must be harvested in the designated year	Eighty-five percent of the grapes must be harvested in the designated year	Both old and new world systems require vintage
Region designation on the label	Hundred percent of the grapes must come from the region designated	When wine is labeled with an AVA, then 85% of grapes within that made the wine must come from there. When wine is labeled with a county, 75% of the grapes must come from there. If wine is labeled with a state, 75% of grape must come from there	If a GI is designated, the grapes must be ≥85% from the GI. Multiple geographical indications acceptable but must be ≥95% and listed in descending order	Both old and new world systems have very high percentage requirements

(continued)

Table 2.7 (continued)

Item	France (old world)	USA (new world)	Australian (new world)	General quality comment
Wine quality classifications on the label	Two categories of wine under AOC system; Vins D'appellation D'Origine Controlee (higher quality, stricter) and Vin de Pays (less quality, less rules). Regions such as Bordeaux and Burgundy indicate their assessment of quality level right on the label	There are three groups of US wines: Varietal, generic, and proprietary. There are no explicit U.S. TTB wine quality classifications or rankings; however, the AVA grape growing region, variety, vintage, and producer combine to provide considerable information to infer the quality of wine	There are no explicit Wine Australia wine quality classifications; however, also similar to US rules, the GI grape growing region, variety, and vintage help infer quality	The French show quality indicator on label base upon quality history and tradition. The new world examples do not

The most famous classification is the 1855 Medoc classification which classified 61 chateaus. Burgundy has its own levels of classification with Grand Cru at the top, Premier Cru second, and Village wines third. Karen MacNeil explained one of the features of the French and European wine producing regions extensive regulations attempt to hold many of the winemaking variables constant within said region (MacNeil, 2015). The land where the grapes grew indicated the powerful quality influence of the terroir. The New World laws and regulation to not attempt to indicate quality, however, provide significant information to enable inferring the quality of the wine.

2.3.3.1   A Closer Look at US Wine Laws, Labels, and Quality

The US wine labeling laws do not explicitly carry a single assessment of quality; however, the rules dictate designations that can be used to infer quality. We are going to explore these requirements further and highlight wine consumer and wine-maker quality interpretations. This will deepen and expand your own quality thoughts and strengthen the quality thread to various wine attributes.

Figure 2.9 is a beautiful wine label annotated with the nine basic wine label requirements.[3] Table 2.8 highlights the consumer vs. winemaker quality aspects. The primary intent is to protect the consumer, but they can and should be noted by winemakers when tasting other wines. The brand name is the producer of the wine

---

[3]The wine label is an original art piece painted by Dona Marino Wieckowski. A Watercolor—Untitled. Provided to us "straight from the heart."

Fig. 2.9 The nine basic US wine labeling requirements collectively paint a quality picture. These are intended for the consumer before purchase and can influence the winemaker in targeting goals for ingredients and finished wine style

Labeling Requirements

1. Brand Name
2. Bottler's name, address
3. Varietal Designation
4. Appellation of Origin
5. Alcohol Content
6. Vintage Date
7. Net Volume
8. Sulfite Declaration
9. Health Warning

and it may carry a history of producing quality wine. This may guide the consumer before purchase or provide a comparison for another winemaker to establish a goal or target of wine style and quality. The bottler is important to ensure safety and long-term stability in the bottle. Varietal designation and appellation of origin are of paramount importance in driving the aroma, taste, color, and other characteristics of the wine. You as a winemaker can make your type and style preferences and seek out comparisons with other producers of your preferred varieties and appellations originating the grapes. Alcohol percentage by volume is controlled by the US TTB, but it is also a feature of how "hot" the wine might taste. The vintage and reviews by wine critics may carry a quality designation and assessment which can be used for comparisons. The vintage by itself, is not sufficient for a unique quality assessment. One year may produce a higher volume but of lesser quality, or another year may produce less volume and still some producers might make a high- or low-quality wine. Keep track of the collective of the producer, variety, appellation of

Table 2.8 Wine label laws are requirements for commercial wine sellers

Item	Description	Comment	Consumer quality aid	Vintner quality aid
1	Brand name	The winery name, the winemaker name, or the wine producer	Names with a history of quality inform consumer before purchase	Set goals, improve brand quality reputation, expand market, and maybe earn a higher price
2	Bottler name and address	The name of the bottler. The city and state must be preceded by "Bottled by" or "Packed by" or other options	Look to the brand name, AVA, variety, and vintage before bottler for quality clues	Important to bottle for stability and safe storage to maintain quality wine
3	Variety, generic, or proprietary class	If designating a varietal type, the label must also show an appellation of origin. If multiple varieties are part of the blend, then the percentage of each grape must be on the label and add up to 100%	Varietal and proprietary groups generally imply higher quality. Generic may still be good but of likely better value	A winemaker can taste and get to know other quality wines with a variety or proprietary blend. This information can help set goals, standards, and guide styles
4	Appellation of Origin (AO) on the label	When a US wine designates a geographical area for the origin of the grapes, it can be a state, county, or an American Viticulture Area (AVA). Federal law requires ≥75% of the grapes must come from within the appellation. States and counties may vary	If the AVA has a history of producing quality wines, chances are the next will be also. A state or county designation with more diverse climates and terroir has is more chance for yearly variation. If the wine is a high-volume value wine, chances are that it will be similar due to blending and vinification practice	Vintner can choose to make wine from known quality growing regions
5	Alcohol content	The alcohol content by volume (ABV) is the percent of the bottled wine that is alcohol	ABV will indicate safe limits, taste, and style	Vintner can target alcohol levels to correspond to a quality balanced wine for given style
6	Vintage date	This is the year the grapes are harvested. If a wine label displays a vintage date, then an AO must also be shown on the label	Historical vintage qualities can imply quality of other wines, many from same vintage grapes	Vintner can track current year growing conditions, compare with prior quality vintages and buy or harvest accordingly

(continued)

Table 2.8 (continued)

Item	Description	Comment	Consumer quality aid	Vintner quality aid
7	Net volume	This is the volume content of the bottle	Reliable indication of how much you are buying	Vintner size final product volumes and intermediary production containers and processing requirements
8	Sulfite declaration	If the total sulfur dioxide or sulfating agent is present at 10 ppm or more, a sulfite declaration must be present on the label	Reliable indicator to consumers with sensitivities	Vintner awareness of legal limits allows sizing of usage quantities and interim testing steps
9	Health warning	Health warning statements are required on any alcoholic beverage with an alcohol content of 0.5% or higher	Reliable health hazard warnings to inform consumer	Vintner can manage harvest and vinification process, and final tests to ensure compliance

They are intended to inform the consumer of quality. They can also be early aids to other winemakers in setting goals and sourcing grapes for future vintages

origin, and vintage to establish a wine style and quality. These four parameters on the label will help identify and locate the same or very similar style and quality bottle of wine.

One more aspect of quality on labels and wine laws within the USA is that each state may tailor some US federal level wine laws. Table 2.9 identifies examples how some state or regional tailoring differs from federal rules and comments on quality implication. This can be done to tighten up restrictions in a direction the state or region intends to highlight quality or special features of their state or region. The following table shows examples of some of California and Oregon state wine labeling laws in contrast to US Federal laws.

2.4   Quality Proxies and How They Correlate

We now have some visual, smelling, and tasting tools to understand the quality of your own wine. We also have some excellent examples of how other wine experts and critics define it. Before we jump into quality cause and effect, let us finish this chapter with a discussion of correlation. This will help us appreciate what it means when various other sources make claims about cause and effect and yet are actually basing their statements upon coincidental correlation. Correlation in itself is not bad, only the act of misrepresenting it as cause and effect. Correlation and understanding correlation patterns can be very helpful while resolving problems and investigating possibilities. Investigating the most probable possibilities is a good

Table 2.9 Each state may tailor certain aspects of US federal wine regulations

Item	Description	U.S. Federal regulation	State or regional tailoring	Quality comment
1	Single varietal designation	• U.S. federal regulation states that at least 75% of grapes used to make a wine must be of the declared variety in an identified appellation of origin	• In the state of Oregon, 90% or more of the wine must be from the named variety. Oregon want to show off its exceptional conditions for growing Pinot Noir	• High concentrations of a grape variety drives higher impact of grape varietal characteristics. This can be a style and quality indicator
4	Appellation of origin (AO)	• Federally, if a wine label lists a country, state, or county as an appellation, at least 75% of the wine must be produced from grapes grown in the place named	• For Oregon wines, 100% of the grapes must be from the state of Oregon • For Oregon labeled AVAs, 95% of the grapes must be from the AVA named • If using the Napa Valley AVA or one of its sub-appellations, 100% of the grapes must be from the Napa Valley AVA	• Regions and AVAs with high-quality grapes, increase the chances of a high-quality wine. Regions want to show off their special characteristic

California and Oregon show enhancements intended to improve their perceptions of quality

way to narrow the universe. Our goal with this section is to understand correlation and know what it means. Our plan is to show some examples, do some simple calculations, and follow with basic interpretations. We will also explore a brilliant wine paper that determines correlations between various important wine quality proxies. All of this will help you as a winemaker set your own quality goals, help you resolve your problems, and guide you on your quest to make quality wine. Yes, enjoy a nice glass of wine before, during, or after this section.

A correlation is a relation between things that may vary with respect to one another that is in a way that is not expected based on chance alone. If two things are uncorrelated, they will vary with respect to each other in a random way. If two things are correlated to some degree, then when one goes up, the other will also go up. When the value of one of two correlated items goes down, then the other will also go down. The degree of correlation is a measure in how tightly coupled these items may be. Philip Bevington wrote an exceptional book (Bevington, 1969) on correlation probability, data reduction, and error analysis.

Let us consider a simple example. Table 2.10 shows an approximate relationship between alcohol concentration and grape sugar concentration. Adding yeast to grape juice is highly correlated with producing wine. This is clear to all winemakers. The yeast will convert sugar in the grape juice to alcohol. This is perhaps the most fundamental process in winemaking. There is also an approximate rough rule of thumb that provides an estimate of the amount of alcohol in the finished wine

Table 2.10 Alcohol concentration (Y) as a function of grape juice sugar concentration at start of fermentation (° Brix)

X (°Brix)	Y (%)
21	11.6
22	12.1
23	12.7
24	13.2
25	13.8
26	14.3

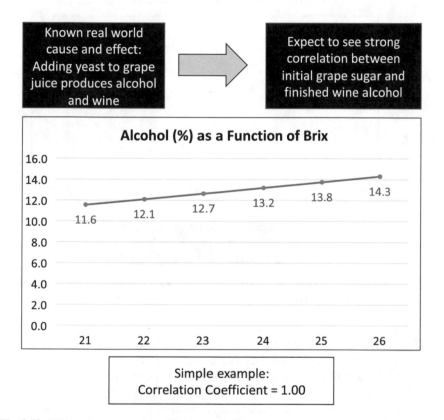

Fig. 2.10 With well-known real-world cause and effect, expect to see strong correlation in data

from the level of sugar in the initial grape juice. A simple math approximate relationship is

$$\text{Alcohol}\left(\%\text{by volume}\right) = 0.55 \times \text{Grape Sugar}\left(^\circ\text{Brix}\right)$$

Where: Brix = grams of sugar per 100 ml of juice.

Let y be the function that represents the alcohol content in the finished wine and let x be the function that represents the sugar concentration in the initial grape juice, then $y = 0.55 \times x$. This relationship is plotted in Fig. 2.10 and results in a correlation coefficient of 1.0.

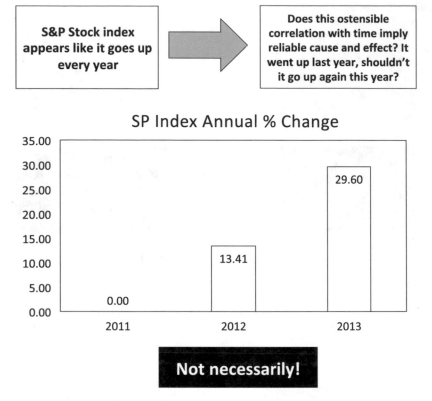

Fig. 2.11 Ostensible data trends may be correlated, but it may not be due to cause and effect and just a coincidence. Quality winemaking requires understanding of cause and effect

Understanding cause and effect is important to knowing what happened and why. It is part of making reliable predictions for future outcomes. If grape quality and winemaking conditions are similar, then achieving a quality wine outcome is likely. Changing input ingredients can cause a predictable effect. Understand the difference between cause and effect vs. random correlation. Figure 2.11 illustrates how a random rise in a stock market price may be a random event.

Establishing a useful quality definition can include associating a handful of attributes or quality proxies that are indicative of a quality wine. These quality proxies may be observable, measurable, and quantifiable. These are the attributes that can help you find your way to making quality wine. At the beginning of your problem-solving investigation, you will not likely understand the cause of your problem. You can narrow the universe of possibilities by looking at certain quality indicators. You can assess your own wine's performance against these quality proxies. Figure 2.12 tries to remind us to ask questions and understand the nature of correlations. There can be strong correlations that will provide exceptional starting points for the winemaker to set quality goals.

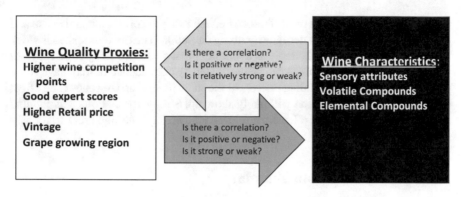

Fig. 2.12 Complex relationships between wine quality proxies and attributes are challenging to navigate

A proxy for quality wine is something that functions, indicates, or represents quality wine. An exceptional article on correlating various quality proxies with wine sensory attributes, volatile compounds, and chemistry is titled "Correlating Wine Quality Indicators to Chemical and Sensory Measurements" (Hopfer et al., 2015). This is not a cause and effect analysis; however, it can help you begin your own definition of quality goals, as well as, focus your search on potential contributing factors. There is always a complex world of variables and factors originating from your grapes or from your vinification process. These may or may not reveal themselves until the finished wine. There will however, always be clues along the way. A thoughtful roadmap of possible correlations can provide a good compass. A reliable link between wine quality, sensory characteristics, and wine components can surely help the winemaker.

We have seen how wine quality means so many different things to different people. We often head back to basics and typically think of quality wine in terms of our own personal preferences on how the wine looks, smells, and tastes to us, how it looks, smells, and tastes to our friends and family, and how it scores in wine competitions evaluated by wine tasting experts. The Hopfer paper makes a broader and deeper assessment of links or correlations between high level quality proxies such as points earned in a competition, wine expert scores, retail price, vintage, and grape growing wine region to sensory attributes, volatile compounds, and wine chemical components. These results can be a useful winemaker tool to help map winemaking ingredients to expected higher quality products.

This study includes controlled sensory tests and a thorough analytical assessment. It provides a clear presentation of correlations between quality attributes, wine characteristics, and chemical components. An effective winemaker can consider ingredients and track quality progress throughout the winemaking process. This will improve the likelihood of you achieving your desired quality outcome.

Previously, we presented a spurious correlation about the stock market, and now we are touting a brilliant paper on correlating wine quality and wine characteristics. We raised the spurious correlation earlier to make sure you have a sufficient level of skepticism and think about the difference between simple correlation and cause and effect. Correlation does not necessarily imply causation. The obvious spurious stock market example is included simply as a reminder that a correlation by itself may not imply any causal effect. Things can be correlated by coincidence or may have some unseen common or confounding effects. At the same time, if driving cause and effect exists, then strong correlation will likely show up. Strong correlations and patterns in large data sets can be an effective guide in establishing links and areas to focus.

2.5 Quality Definition Summary

Establish and understand your own definition of quality. Set your quality plans early in the winemaking process. This provides a roadmap for making your wine, markers to measure progress, and checkpoints to know when you have arrived. This is one of many tools you can use to avoid, mitigate, and resolve problems you may encounter. We mentioned many times how important it is to develop your own good sense of what a quality wine looks like, smells like, and tastes like. One of the aims of this book is to teach winemakers how to set goals, make quality wine, and solve winemaking problems. A clear and concise quality definition is essential to begin your journey of making quality wine. We do not wish problems on you; however, even with the best intentions, we expect you may encounter them. On this problematic road of making wine, it is always good to know where you are headed. Read on, we will share some powerful preventive methods to avoid many problems along with insightful tools to resolve them when they do show up. Alas, nature does not always cooperate.

The vineyard is the place to start setting your quality goals. Create your own quality goals and understand what they mean. Goals may have different aspects and layers. Be clear and specific. Goals for this year's harvest are timely, but do not forget to look 2, 3, and 4 years out. Be clear on details. Set your quality goals before the growing season starts, prior to bringing the harvest into the winery, and long before the wine goes out the cellar door.

References

Alcohol and Tobacco Tax and Trade Bureau, US Department of Treasury. (2019). *Alcohol regulations*. Retrieved from TTB: Alcohol and Tobacco Tax and Trade Bureau: https://www.ttb.gov/other/regulations.shtml#alcohol

Amerine, M., Roessler, E., & Fillipello, F. (1959). Modern sensory methods for evaluating Wine. *Hilgardia, 28*(18), 477–567.

Amerine, M. A., & Winkler, A. J. (1944). Composition and quality of musts and wines of California grapes. *Hilgardia: A Journal of Agriculture Science, 15*(6), 493–675.

Australian Wine, T. A. (2019). *Growing and making, labeling*. Retrieved from Wine Australia: https://www.wineaustralia.com/

Bevington, P. R. (1969). *Data reduction and error analysis for the physical sciences*. New York: McGraw-Hill Book Company.

Hopfer, H., Nelson, J., Ebeler, S., & Heymann, H. (2015). Correlating wine quality indicators to chemical and sensory measurements. *Molecules, 20*, 8453–8483.

Johnson, H., Robinson, J., & Beazley, M. (2007). *The world atlas of Wine* (6th ed.). London: Octopus Publishing Group.

MacNeil, K. (2015). *The Wine bible* (2nd ed.). New York: Workman Publishing.

Parker, R., Jr., & Rovani, P.-A. (2002). *Parker's wine buyer's guide* (6th ed.). New York: Simon & Schuster.

Robinson, J. (2019). *Learn about Wine*. Retrieved from Jancisrobinson.com: https://www.jancis-robinson.com/

TTB Alcohol and Tobacco Tax and Trade Bureau. (2019). *Sulfite mandatory labeling items*. Retrieved from TTB.gov: https://www.ttb.gov/wine-resource-tool/section07-labeling/sub-section1/71120.htm

Chapter 3
Root Cause Analysis Applied to Grape Growing and Winemaking

Often when we have a problem with wine, we think of the most obvious cause of the problem and move on. But winemaking has such long periods of time between processes and before the same procedures are repeated that you will not know for a long while whether you have taken the proper corrective action to prevent the problem from recurring. In the meanwhile, you have processed more batches of wine with potentially the same mistakes repeated. Say, for example, you have a high volatile acidity problem with a barrel of wine and you "assume" that it was due to not topping off the barrel, thus allowing oxygen to promote the growth of spoilage microbes. Without also considering a lack of moldy grape sorting, poor sanitation practices, and infrequent sulfur dioxide additions, the problem of high VA may happen again the next year. If the problem was with a single barrel vs. multiple ones, this offers some clues, as well as whether there was a single vs. multiple employees involved with the operation. Root cause analysis is a set of methodologies to help identify all possible causes of a given problem and then determine which one is the most likely root cause of that problem. Our RCA approach isolates contributing factors.

3.1 The Cause and Effect Diagram (The Fishbone Diagram)

The cause and effect diagram (George, 2002), also known as the fishbone diagram, and the Ishikawa diagram, lays out six major categories of problem causes for any given problem effect. You use the diagram in Fig. 3.1 by writing the effect of a problem at the far right of the fishbone and then consider causes in the categories of people, processes, equipment, materials, measurements, and environment. For example, if the effect is a wine that smells like hydrogen sulfide (rotten eggs), first you would write that at the far right of the fishbone (the head of the fish.) Then you look back in time at all the processes performed up to that point in time, all people

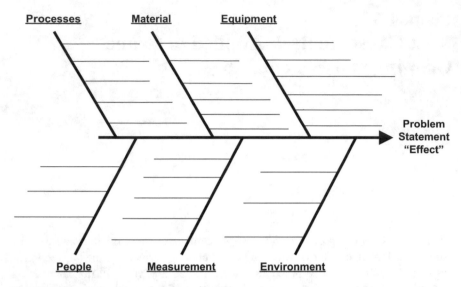

Fig. 3.1 Generic cause and effect (fishbone) diagram

who have been involved so far, all the equipment used, the materials used, the measurements performed, and the environments you have been in up to that point. In this example you would identify potential root causes including any high temperatures during fermentation, use of wild yeast that may overproduce sulfides, not using a yeast food at all or measuring out too little food, use of a domestic yeast that is prone to H_2S production, mismeasurement of the amount of yeast needed for fermentation, untrained people preparing the yeast inoculation, skipping a process step like not measuring the temperature for the yeast hydration, etc. This procedure is helpful to perform with all the winery staff using a brainstorming process where you collect all the possible causes without having any detailed discussion or passing any judgement yet. You want everyone's mind open to bring up all possible things without trivializing them. Some ideas may not be good, but hearing them may trigger a better idea which you otherwise would not hear. You may also discover that one employee performs a process different from everyone else, which will be vital to the investigation. Once the causes are all collected and documented on the fishbone diagram, then you can go through each one systematically to rule out any that are not applicable or not valid.

In Fig. 3.2 you see a fishbone diagram designed for winemaking. This can be used to jog your memory as you work out a specific root cause analysis. For example, in the equipment category we have listed your destemmer/crusher, press, tanks, barrels, and bottling line. But it may also be necessary to consider the forklift, sorting table, conveyor, hopper, macro bins, punch down tool, pump over cart, pumps, hoses, topping cart, steam cleaner or CO_2 cleaner, racking wands, sampling equipment, bottle cleaner, etc. The material category is different in that it does not include any machinery, but it is all the materials that get added to wine or come in contact

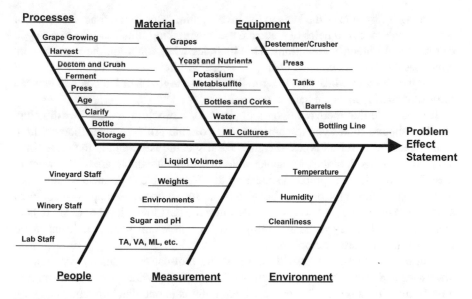

Fig. 3.2 Winemaking cause and effect (fishbone) diagram

with wine such as cleaning and sanitizing solutions, the grapes themselves, yeast and associated nutrients, water, malolactic fermentation cultures, and various additives including potassium metabisulfite, tartaric acid, enzymes, color enhancers, stabilizers, etc., as well as the bottles and corks.

In the processes category, these are the step-by-step methods that get performed to make wine. You may be using certain equipment and materials from the other two categories but this category involves the actions you are taking such as the steps for hydrating yeast for inoculation, or the steps to clean barrels. The fishbone diagram helps you to separate these elements to determine whether the barrel cleaning equipment is faulty, or the water or cleaner was bad, or a step in the process was omitted or performed for the wrong amount of time. Similarly, when you consider the environment category, you evaluate whether the problem only occurred under certain temperature or humidity conditions, or inside vs. outside, etc. For the measurement category, you need to evaluate all measurements that, if incorrect, could be the root cause of a problem. An example would be if the scale for weighing incoming grapes is significantly off, then the amount of potassium metabisulfite you add at crush may be too much or too little, or the yeast added for primary fermentation may be too little or too much. In the people category, you consider all the people involved and whether they are the source of the problem. You may have good equipment, good materials, a well-documented and correct process, a stable environment, and calibrated test equipment, but one employee performs the process by memory and was poorly trained, so the problem only shows up when she or he performs the task. There may be a tasting room employee who does some periodic barrel tastings with club members but forgets to let the winery staff know exactly when and which barrel was involved.

Timing is also one of the keys to the investigation. Relate the time and phase of when changes occur. It helps to associate the winemaking status and process so you can accurately identify conditions. This includes what condition the wine was in, what was done to it, and what changed. These will help provide the clues to what caused the problem. Make sure to write it down in a logbook or lab notes so you can retrieve it later. Details are hard to remember.

It is possible that there are two root causes for a problem you are investigating, but the fishbone forces you to consider each possible problem independently first. The intent is to cover all types of causes for a specific issue and at the same time, simplify the considerations to manageable and understandable areas. We want to identify and explain the problem so clearly that we are able to define a solution that will not only resolve the problem but also make sure it does not happen again. Each successive step in the process involves defining the next lower level of detail to help isolate the problem. By virtue of starting with the whole universe, we help ensure no probable suspects are overlooked.

Cause and effect analysis will help find the root cause of quality problems in your wine and minimize lost production of poor quality wines. Cork taint can ruin a beautiful wine. Figure 3.3 shows a winemaking fishbone diagram annotated with questions regarding potential cork taint issues. It has its origins in poor quality corks and the chemical compound, TCA (2,4,6-trichloroanisole) (Boulton, Singleton, Bisson, & Kunkee, 2010). It can result in musty or moldy newspaper smells. It is

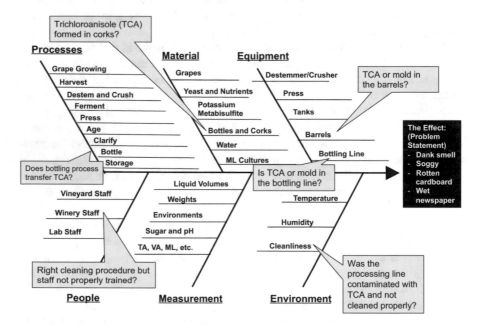

Fig. 3.3 Cork taint produces a striking wine flaw. Cork taint is caused by the presence of a chemical compound called TCA. However, what is the root cause for its presence in your wine? Simply replacing all your corks or closure type may be premature. Understand the cause, so you can fix it. If not this wine, next time

created by reactions between plant compounds and mold and can originate or be transferred through your corks. Should you just throw out the bad batch of corks and try again? It may have originated in the corks, but where is it now? Investigate your complete winemaking operation with a cause and effect assessment of your bottling process, corks, your aging barrels, your bottling equipment, the training of your winery staff, your cleanliness procedures, and the actual cleanliness of your equipment. A careful and thorough cause and effect analysis will take time; however, it may salvage next year's vintage. It will save you time and money in the long run.

3.2 The 5 Whys

The 5 Whys is a simple method to get to the root cause of a problem without requiring any complex statistical analysis. Instead of asking once why something happened, you ask why up to five times to understand what really needs to be fixed. Here is an example of why you do not stop after asking why just one time:

In this example we go beyond blaming Joe, to determining that a written procedure and proper training will improve the probability that everyone will perform this process in the same manner in the future with a high degree of repeatability. Figure 3.4 shows example questions and answers. And we see that it only took 4 Whys to determine a probable root cause, which is OK! The 5 Whys method is especially effective when human interactions are a factor, when evaluating multiple root causes, and when combined in conjunction with the fishbone diagram.

3.2.1 Another Set of Penetrating Questions

Another simple line of questions helps to sort through complex situations and penetrate to the root cause. These were suggested in a fascinating book about investigation and explaining failures in airplanes (Soucie & Cheek, 2011). Airplanes and

Questions	Answers
1. Why did this wine get over-sulfited?	Joe did it
2. Why did Joe make a mistake?	Joe has a bad memory
3. Why does Joe need a good memory?	The procedure for sulfite addition isn't written down
4. Why does the procedure need to be written down?	There are four employees who perform this task and it isn't performed frequently.
5.	

Fig. 3.4 Keep asking "Why" until you get to the true root cause of the problem

wine have very little similarity; however, investigating quality problems and accidents has some overlap in the type of questions that should be asked. These two have been distilled from David Soucie's analysis. They are designed to fit our wine quality problem discussion.

1. Was grape growing and winemaking information available that could have prevented the quality problem?
2. If the information was available, did the right people (your vineyard management team and winery staff) have it at the right time and were they knowledgeable and trained to recognize and act upon it?

These are a good set of overarching questions to guide your own quest for improving the quality of your wine. Given that so many quality wines are now produced around the world, the answer to the first is almost always yes. However, note there are millions of different growing and winemaking conditions, they change every vintage, they change every month, and they change from winemaker to winemaker. There is a chance that your particular situation and quality issue may be unique. In many cases, the proper grape growing and winemaking information is available; however, your vineyard manager or winemaker may not be informed soon enough. Preventive measures for improving the quality of your wine are very effective, as long as they are applied at the right time. You are counting on your team to tend the vineyard and make the wine. Their knowledge, training, and motivations need the same quality attention as the making of the wine itself.

3.3   The Kepner–Tregoe Analysis Method: Focus on Decision-Making

There is a well-known and proven method for rational decision-making that can be applied to winemaking problem-solving. This is the Kepner–Tregoe analysis (KTA) method that was developed in the 1950s in the USA by Dr. Charles Kepner and Dr. Benjamin Tregoe (Kepner & Tregoe, 2013). This is a logical approach on how to think about a problem and its solutions. There are many historical and recent examples of how it was effectively applied to complex problem-solving and decision-making. We all appreciate that there is tremendous joy that can be found in a quality wine. Wine is very personal and depends so much on individual tastes. There is often skepticism about applying a rational logical process to something that is so sensuous, fun, and creative. Well, we believe this rational method can truly be applied to such a thing, and that is of course, winemaking. It can and will improve the quality of your wine. In this section, we will outline the basic approach and in other sections, we will present winemaking problem scenarios. There are many aspects to the many KTA methods; however, we are going to build upon the decision-making portions and show how the process can be applied to aid in better decisions and improving the quality of your own wine. This will occur after you have done some basic groundwork on facing your problem with essential situation assessment, basic cause and effect analysis, and are ready to explore alternative options to fix and mitigate the problem.

Let us first review some of the essential features and benefits and then layout the basic mechanics of the process. Table 3.1 summarizes the features and benefits of the KTA approach. Using a rational problem-solving approach will lead to a rational decision. It is almost self-evident and will not lessen the joyous wonders of the wine created. It is a key part of this approach to first assess and document the status of the situation. This may be a part of every efficient and effective problem-solving approach. This should take a little time before you embark on a recovery plan. This will improve your chances of not wasting time on solving the wrong problem.

Thinking about options and alternatives to fix your situation also takes work and knowledge. In most occasions, you will not only have to outsmart your problem, you will have to outwork it. By thinking about the pros and cons of problem resolutions, you will identify not just the likely positive outcomes, but also the possible real-world adverse consequences or risks. Considering both leads to a better-balanced decision. This integrated numerical approach enables a rational approach to compare options. Let us now break down the steps to implement this KT analysis and decision-making approach.

Table 3.1 Features and benefits of the KTA approach

Item	Feature	Benefit
1	Rational approach	Provides a logical path to make an informed and optimal wine problem resolution decision. This leads to the most sensible choice
2	Assess and clarify the problem situation	First get a good understanding of the problem and associated characteristics. Understand the details before your waste precious time and energy on solving the wrong wine problem or path to fix
3	Explores alternatives	This broadens the world of possibilities to improve your chances that you are considering the most prominent wine issues, probable causes, and resolution options
4	Evaluate pros and cons of the situation	Provides better clarity to focus on improving what is good and mitigating or removing what is bad
5	Weight the characteristics	This quantifies and allows proportional balance to what is important in evaluating the wine problem resolution options that might be available
6	Score the options	Provides a quantitative score in accordance with your own wine preferences and situation constraints to allow ranking and evaluating the severity of the characteristics you are considering
7	Evaluation of risks and negative aspects	Includes evaluation of the risks due to uncertainties and negative aspects of unwanted consequences. Uncertainty, risks and potential negatives aspects are a part of all imperfect real-world situations and actions. It helps to "fold" these into your assessments
8	Integrated assessment of options	Integrated evaluation incorporates your own personal and team weights of what you like and dislike. It also includes the down sides of unwanted consequences. This approach results in the best choice which is a balanced selection of what you consider good and bad for this situation

These are the general steps that we like to use to implement and apply this KT approach to winemaking problem-solving.

Table 3.2 provides an orderly sequence to proceed through this rational decision-making process. Of course, the order may be shuffled or approximated as you work this approach, get familiar, and tailor to your own preferences and what works best for you. We make sure you begin with a good assessment of the situation and conditions of your problem. Then follow with identifying your requirements or "must haves" and objectives or "wants." Clarity and specificity help you work the accuracy of your situation. These are the positive attributes and characteristics you want to achieve in the resolution of your problem. Assign a relative weight to each in accordance with your own values and degree in which you want these positive results. Assign the highest numerical weights for those items that are most important to you.

You will also have to work to identify various options or alternatives that might repair and/or prevent this problem from happening again. Evaluate each option against each positive quality in terms of how well you think it will accomplish or satisfy the objective of your positive characteristic. These scores should range from 1 to 10 for the least to the most likely to be successful. Consider each and every

Table 3.2 Steps to implement KT analysis

Step	Description	Comment
1	Assess the situation and document the problem	• Improves your understanding of current conditions and accuracy to begin problem-solving
2	List attributes or requirements and objectives and weight them on their importance	• Weight them by how important they are to you and your goals. Bigger weights are more important. Weights can be on a scale of 1–10 or can be higher if the importance is significantly higher than the other attributes you have chosen
3	Identify options or alternative actions to resolve or prevent problem	• Create a big plausible list of alternative actions that might resolve your problem, now or in the future. Think broadly now, we will narrow in subsequent steps • Do not include those that do not meet your minimum requirement threshold
4	Score each option or alternative from 1 to 10 on its ability to satisfy your objective and calculate the weighted score	• The weighted score is the product of the objective weight times its raw score • This numerical evaluation allows us to readily compare options
5	For the top options, identify and list negative adverse consequences	• There will be unwanted consequences of each alternative that we will consider in making a balanced decision
6	For each adverse consequence, estimate probability of occurrence and weight by severity of impact	• This is like above numerical evaluation of positive effects, but in the negative or unwanted world
7	Choose the best option by picking the one with the highest positive characteristics and least adverse effects	• This offers the best and most balanced decision, as defined by your own criteria

option, one at a time, against each of the attributes, one at a time. The absolute values are not critical, but the relative score is. You may even make sure the most likely gets a 10 and the least likely a 1 and the rest can be spread in the middle, with respect to the highest and lowest. It may not be easy to discriminate or assign a value. It is also important to look across and between the various options and consider one particular characteristic. Consider the relative scoring between the options for a given attribute and assign a higher score to the one most likely and the lowest score to the one least likely. It is the relative ranking that is important, not the absolute value.

Figure 3.5 provides a graphical display and summary of an evaluation and assessment of a problem with three possible resolution options and three wanted quality attributes. Each quality attribute gets its own row. Note that in this example, there are three attributes. For each attribute, you will input the numerical value of the weight you would like to assign to each. In this example, our highest weight is 20, the middle is 10, and the lowest is 2. Create the spreadsheet to calculate the fractional weight of each by dividing by the sum of all the weights. In this case it is 32, making the fraction weight of the first attribute equal to 20/32 or 0.625 or 62.5%. You will also input the option score for each of the respective attributes. Note that for option 1, we have assigned a score of 10 for achieving the first quality attribute. The spreadsheet calculates the product of the fractional weight times the score to give a weighted score of 6.25. This is similarly done for all the options and for each of the respective attributes. The integrated weighted score is computed by summing the weighted scores. The best choice of the various options is the one with the highest score.

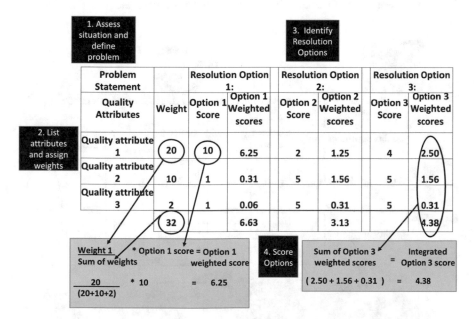

Fig. 3.5 Integrated KTA decision matrix displays quality problem, quality attributes, resolution options, and scores using normalization formulas for integrated score

Note this method can extend to a larger number of options. You may add more columns, but remember this requires more research and work. A few top promising recovery options may be sufficient. You may also add more quality attributes or rows. Here again, this is up to you to decide how many you want to pursue. A few top key attributes may provide what you really need and want and will be more manageable than a larger number of quality attributes. The corresponding steps that correlate with our preferred approach are shown. These certainly may be reordered, after you gain familiarity and establish your own preferred techniques. Figure out what works best for you, but preserve the delineation of options, attributes, weights, and scores. We will provide an alternative problem-solving approach ("Quick-look") that is rational, faster, but with a little more risk, later in the book. One size does not fit all.

Some people like to think in terms of percentages. Instead of assigning numerical integer weights and normalizing, one could directly assign percentages to each weight. Using percentages, make sure the sum of the various attributes adds up to 100%. One might want to have quality be the biggest driver and get 90%. All the other attributes need to add up to 10% to make the total 100%. There might be another case where one quality attribute like color represents 1/3 of the decision, aroma represents another 1/3, and flavor represents another 1/3. In this case each of these three attributes would be assigned 33.3% to add up to 100%. This represents everything that is of value in this decision. Sometimes, using fractions or percentages is a more direct way to translate your thoughts of relative importance into weights. The KTA decision matrix in Fig. 3.6 uses the same relative weights as Fig. 3.5, except it is mathematically populated with percentages.

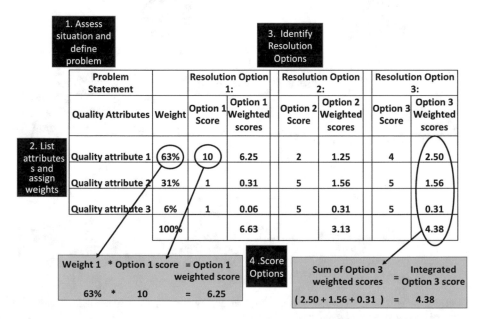

Fig. 3.6 Integrated KTA decision matrix using percentages for weights that add up to 100%

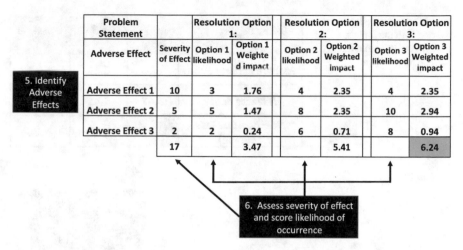

	Problem Statement		Resolution Option 1:		Resolution Option 2:		Resolution Option 3:	
	Adverse Effect	Severity of Effect	Option 1 likelihood	Option 1 Weighted impact	Option 2 likelihood	Option 2 Weighted impact	Option 3 likelihood	Option 3 Weighted impact
5. Identify Adverse Effects	Adverse Effect 1	10	3	1.76	4	2.35	4	2.35
	Adverse Effect 2	5	5	1.47	8	2.35	10	2.94
	Adverse Effect 3	2	2	0.24	6	0.71	8	0.94
		17		3.47		5.41		6.24

6. Assess severity of effect and score likelihood of occurrence

Fig. 3.7 Adversity matrix displays adverse effects, severity of effects, and likelihood of occurrence

This KTA matrix covers the evaluation of positive attributes for each option. Let us now move onto potential adverse effects. Figure 3.7 shows the matrix structure of how this is implemented. These are not desired but often come with real-world solutions. Just about all problems require some level of cost and schedule to resolve and these are classic adverse effects. More is considered worse. Some wine chemistry and environmental treatments will have multiple outcomes and uncertainty. Some of these effects will make some of our quality attributes better; however, many adverse effects can also show up. The potential adverse effects are identified and listed as part of step 5. One needs to think about the severity of these adverse effects given the status of your own situation, requirements, and constraints. In addition, the likelihood of them materializing or occurring is also part of step 6. The product of their respective severity weights and likelihood of occurrence is their weighted impact or adversity score.

The last step in this KT analysis or decision-making method (step 7) is to subtract the adversity score from the positive integrated attribute weighted score. Figure 3.8 shows the three options and the sum of their positive quality scores and tabulates their adverse score. The option with the greatest value is the rational choice. This reflects the status of your problem, your own quality attributes, your own relative values, and your own assessment of the situation. This is a rational decision-making process that can be applied to your winemaking.

Fig. 3.8 Best decision is
the one with the highest
positive qualities and the
least adverse effects

Balanced Decision	Option 1	Option 2	Option 3
Positive Quality Score	7.56	3.13	4.38
Adverse Score	3.47	5.41	6.24
Net (positive less adverse)	4.09	-2.29	-1.86

7. Best balanced
decision gets the
largest net score

3.4 The Best of Best and Worst of Worst

Another root cause analysis tool is called "best of best and worst of worst" or
"BowWow"[1] for short. The world of winemaking is big and wonderful which unfortunately also means you will likely face problems that are often very complex to
solve. BowWow helps find plausible suspects for the cause of your problem. It looks
at the link between the best and worst finished wines and the best and worst ingredients, equipment, or other vinification aspects. The simple premise is that often, the
best components will make the best wine and the worst components will make the
worst wine. This is a great way to start identifying good suspects. It may not be
always true; however, it improves your chances of identifying problem suspects.

This technique has no specific template or structure to follow. You need to pour
over data to find the best and worst conditions for a particular situation, and then
display the results in any way that allows you to interpret them best—tables,
spreadsheets, diagrams, plots, pictures, etc.,—all are acceptable ways to analyze
the data.

---

[1] This process was given a pet name, the "BowWow" method, by Dr. Howard Sawhill. This was
shared with us over a quality glass of wine and inspiring conversation in 2018.

References

Boulton, R. B., Singleton, V. L., Bisson, L. F., & Kunkee, R. E. (2010). *Principles and practices of winemaking*. Davis: Springer.

George, M. L. (2002). *Lean six sigma: Combining six sigma quality with lean speed*. New York: McGraw-Hill.

Kepner, C. H., & Tregoe, B. B. (2013). *The new rational manager: An updated edition for a new world*. Princeton: Princeton Research Press.

Soucie, D., & Cheek, O. (2011). *Why planes crash, an accident investigator's fight for safe skies*. New York: Skyhorse Publishing.

Chapter 4
Quality Grape Growing and Winemaking, Nurtured with Family, Friends, and Fun

Let us begin with establishing a standard winemaking process that we can use as a common frame of reference. Even though every winemaker is unique and creative with individual style, there are certain fundamentals that will represent a common foundation. Figure 4.1 shows the winemaking process starts with growing grapes and ends with storing bottles. We will use this standard 9 step sequence to frame our discussion. Isolating the problem to its respective phase of occurrence helps narrow the possible suspects.

4.1 Step 1: Grow

When you start your vineyard from scratch it is important to test your soil and understand your grape growing region. Buy the appropriate grape varietal and rootstock that is certified against phylloxera. In Fig. 4.2 we chose rootstock suitable for our clay loam soil and with moderate vigor.

Some of us are grape growers while others are winemakers, and some of us are both; however, all of our wine starts with the quality of the grapes as they come from the vineyards. We may inadvertently make bad wine from good grapes but we will never be able to make good wine from bad grapes. Whether you buy grapes, grow your own grapes, plan to buy vineyards, or plant your own vineyards, understanding the quality of the vineyard is a powerful early step to making great wine. The terroir, soil, sun, rain, temperature, and integrated pest management all contribute to the health of the vines and quality of the grapes.

We are located in the Santa Cruz Mountains in California and are in the vicinity of a number of vineyards with a long history of quality vineyards and wines. Figure 4.3 shows our hillside vineyard early in the season in its third year. Working with an established grower of quality grapes used to produce quality wines is one great way to start.

© The Editor(s) (if applicable) and The Author(s), under exclusive license to
Springer Nature Switzerland AG 2020
J. Steakley, B. Steakley, *A Quest for Quality Wine, Every Time*,
https://doi.org/10.1007/978-3-030-34000-1_4

Fig. 4.1 Nine basic winemaking steps

We want high yield to get the most product and sales from our vineyard. We want minimum pruning weight to minimize cost of pruning and maximize profit. We want the minimum number of vines to minimize cost of planting and maintenance of the vines. We also want best balance to not under or over-crop and provide for the most sustainable vineyard year after year.

Work to understand the suitability of your vineyard's type for your particular varietal type. We are growing Bordeaux varietals in soil that is clay loam which is acceptable, but does not provide optimal drainage. We have noticed in our own hillside vineyard that the highest yielding vines are also rows with the most hours of sun per day and the best drainage. Figure 4.4 describes the layout of our vineyard. The row spacing should be large enough for maintenance and harvesting equipment but close enough to provide high yields per acre. Though our row spacing is 8 ft, the hillside is too steep to drive a tractor through.

Fig. 4.2 Our first vineyard set of varieties were grafted with moderate vigor, phylloxera-resistant rootstock

The in-row spacing should not be so tight for vines to compete with each other for nutrients and close enough for good yields. The actual dimensions may vary vineyard to vineyard, but our vineyard is typical with about 8 ft between rows and 5 ft in-row spacing. This gives each vine about 40 ft^2 of root access per vine.

Our rows are North South and we are located on a hillside facing west, which is not an optimal sun angle but it is acceptable. A Southern exposure would be better. Our trellising is the VSP or vertical shoot positioned approach. In general, grape-vines benefit from around 7 h of sun per day. A Southern facing exposure would be better; however, our grapes do ripen with warm summer temperatures that help make up for this sun angle deficit.

We chose phylloxera-resistant root stock. Let us review what is important for rootstock selection. We want good resistance to pests and disease. We want a good match to the soil texture, depth, fertility, and chemistry. We want the rootstock to be compatible with the water availability and irrigation practices. We also want it to be appropriate for the fruiting variety.

Growth rates of plants increase with temperature. We use an on-site weather station to collect vineyard environmental data. Growing degree days is a summation of daily averages of the max and min temperatures above 50 °F. Know the temperature profiles or climatology for your location. A history of temperatures should match the respective varietals you are expecting to harvest.

Cool night and warm day diurnal variations can also be important to the quality of the grapes. The grape vines need water, whether it is from irrigation or rainfall.

Fig. 4.3 Hillside vineyards have drainage and aesthetic advantages but bring challenges

We have a drip irrigation system that helps keep the plants alive and hopefully stressed. A minimal amount of water keeps the berries small resulting in greater concentrations leading to better tasting wine. California has been in drought conditions for many years and this has cut our yield by close to half, but were expecting the quality to remain high.

Canopy management is a must. We employ a Vertical-Shoot-Profile approach to trellising. This requires wire training early and periodically during the season. We nominally try to achieve two clusters per shoot and two shoots per spur and one spur on the cordon every 6 in. Figure 4.5 lists a few vineyard tips we have found useful in vineyard management.

Pest control is also important. We have been plagued by powdery mildew. Given the heavy powdery mildew pressure in our location, we use a high-pressure sprayer shown in Fig. 4.6. We manage it by spraying near weekly with an organic mineral oil right before bud break and all the way up to about a month before harvest. Thinning the vineyard to allow airflow between shoots and grape clusters is also important to improve spraying effectiveness.

We keep the many local deer out of the vineyard with an 8-ft-high perimeter.

We have a light cover crop of mostly natural grasses but mow and trim to keep it down and not compete with the grape vines. If you are a grower, walk the vineyard near daily throughout the season to monitor and address changes. Disc the cover crop into the soil in the Spring if possible.

Fig. 4.4 Eight feet between rows and 5 ft in-row vine spacing supports access and moderate yield

Fig. 4.5 Vineyard tips

Vineyard Tips:

1. Start shoot thinning early in April, May, and June to avoid energy going to shoots and away from grape development
2. Leaf pull in June, July, and August to minimize vegetative flavors and also submit petiole plant material for analysis and assessment of nutrients
3. With temperatures cool in the morning and warmer in the afternoon, leaf pulling on morning sunny side of the vines can aid in berry development

The birds fly around smelling the sugar in the grapes as they ripen. If you have a very large vineyard, you may choose not to manage the birds since there are far more grapes than birds. Some vintners put up mechanical birds of prey or even use falconry to deter birds from robbing grapes. As shown in Fig. 4.7, we use 12-ft-wide bird netting to cover both sides of the vines when the birds first show up and do not remove them until after harvest. We use an optical refractometer to measure sugar content but take an early cue from the birds in August. The shiny ribbons did not seem to scare them off.

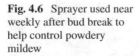

Fig. 4.6 Sprayer used near weekly after bud break to help control powdery mildew

Managing the rodent population is also important, especially gophers and moles that can tunnel through plant roots and kill vines. Trapping is effective but with large vineyards this can be very labor intensive and costly. We have two owl boxes that are frequented by screech owls who are natural predators of rodents. As shown in Fig. 4.8, we also have bobcats which are great hunters.

A healthy, thriving, and well-managed vineyard is a true indicator of quality grapes. We have accomplished the first set of prerequisites. Now let us move on to perhaps the most important decision in winemaking, when to pick the grapes.

4.2   Step 2: Harvest

The picking day is one of the most important decisions in winemaking. Once you have picked, the grapes are set, and the clock starts ticking. On the road to preparing to pick, frequently look at the plants, taste the berries, and test the berries. Figure 4.9 shows an optical refractometer for testing the sugar level in the field. Make sure the portion you test is a good representative sample of the multiple clusters, canopy locations, and vineyard sites. Check the weather, avoid the rain, plan for a picking

Fig. 4.7 The birds are present in late Summer; however, they do not start eating the berries until they are almost fully ripened. Bird nets will protect the grapes

Fig. 4.8 Vineyard cats, even local bobcats, can be effective hunters to keep the rodent, gopher, and mole population down

crew, arrange for destemming/crushing equipment, and fermenting tanks. Make provisions for different varietals and different vineyards with unique climates ripening on different days. If this is your first season, start a written record of events throughout the harvest, if you have done this before, start planning with a review of events from prior years.

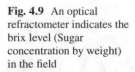

Fig. 4.9 An optical refractometer indicates the brix level (Sugar concentration by weight) in the field

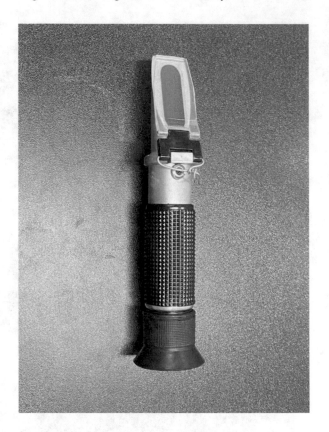

Note when the plant stems and shoots start changing from green to brown. The leaves will change to autumn shades of yellow, red, and brown. Many of the leaves will begin to fall. The berries should be nice and plump. Taste the fruit and it should be juicy and flavorful. Later in the season, the berries may have started to shrivel and look like raisins but still may have adequate juice and be suitable for winemaking. There should be little to no green in the pulp. Figure 4.10 shows our Pinot Noir on the vine. Typically, we harvest our Pinot Noir in mid-September, Merlot and Malbec in late-September, Cabernet Sauvignon and Cabernet Franc in early October, and Petit Verdot in mid to late-October.

Inspect the seeds and taste the skins. Historically, green seeds may impart too many tannins which could contribute too much astringency and bitterness. The seeds should be half to mostly brown. The skins contain aroma and flavor elements. The berry and skin flavor and taste help indicate ripeness and readiness to harvest. Also, extended maceration is also key to bring in the right amount of color, aroma, and flavor effects.

Figure 4.11 shows a few harvest tips we have been collecting. Start your own list of tips.

Establish your own preference for taste and flavor, because probably every grape grower and winemaker has a different opinion. In many ways, sugar, pH, and

Fig. 4.10 Grape varietals
ripen at different rates

Fig. 4.11 Harvest tips

Harvest TIPs:

1. Develop harvest plan and list of grape
   characteristic indicators at least a month
   before target picking date (for example:
   color, taste, sugar, and acid)
2. Understand grape indicators impact on
   wine and check tastes and values weekly
   before target date and near daily as close
   in. Keep track of grape indicator trends,
   weather, and plan ahead for the best
   estimate of the picking date
3. Provide careful instructions and
   examples of good fruit vs bad fruit
   clusters for the field crew to kick off the
   actual picking

titratable acid are good indicators because they have quantitative measures.
However, a great winemaker and viticulturist also knows the taste of ready-to-
pick grapes.

Be outgoing and talk to your neighbors and note when they are picking. Given
close proximity, it is likely there is some similarity in climate, soils, and varietals.
Sharing information can help the entire community.

Fig. 4.12 If possible, get out early before the sun comes up. There will be a slightly higher sugar content in the grapes before the plant gets warm. This will also be cooler for the picking crew

Brix will increase as the grapes ripen and tends to flatten out toward the end of the season. By waiting later into the growing season, the higher sugar will produce a higher alcohol concentration in the wine. But also note that as the season progresses, the acid will be decreasing and the pH will be increasing.

For many years, sugar concentration or sweetness was the primary deciding factor. Be careful to get broad and representative samples from all parts of the vineyard intended to be harvested. For example, pick sample grapes from the top of the clusters,

Fig. 4.13 Sharpened and cleaned cutters will increase efficiency and limit propagation of viruses

bottom of the clusters, inside the canopy, outside the canopy, from vines at the top of the hill, and vines at the bottom of the hill. This minimizes errors from testing over or under ripen grapes compared with the rest of the vineyard. Observing and testing for multiple indicators can provide a more balanced picking date that better fits your needs. Evaluate all of your considerations and note the trends to assess the impact on your preferred style of wine. Later in the season, there will be changes in the grape characteristics, but they will likely be small and slow as temperatures drop.

Except for in the rare cases of extreme weather storms, it is expected you will have multiple days around a "good harvest" window. There really is no simple good formula for the picking date, but at least you can establish a list of criteria that helps fit your own situation and stylistic preferences. If you are lucky, as seen in Fig. 4.12, you might get a nice cool morning that will be easy on your crew.

Be sure all tools are clean and sharp, see Fig. 4.13.

Having a crew ready and available to pick can be challenging. Weather forecasts can drive the urgency to harvest and can impact the availability of your crew. We

Fig. 4.14 You have stressed the grapes, which is a good thing, now tailor this generic harvest checklist for your own circumstances and execute early to reduce stress on the winemaker and vineyard manager

> **Generic Harvest Checklist**
>
> **Vineyard Picking Site:**
>
> ☐ Picking crew
> ☐ Picking bins
> ☐ Grape vine cutting knives, hats, and sunscreen
> ☐ Hand washing soap, buckets, towels
> ☐ Refractometer kit
> ☐ First Aid Kit
> ☐ Water, Coffee, and Donuts for crew
> ☐ Pop-up shade tents
> ☐ Restroom facilities
> ☐ Log Book
> ☐ Transportation from vineyard to crush-pad and fermenting tanks

solicit family and friends to help with this challenge. Since different varietals will be ready for harvest on different days, this helps spread out the workload.

Figure 4.14 shows our harvest checklist. Make your own checklist and pay close attention in the weeks preceding target days. If you have made wine from these grapes in the past, you will be able to use this data and work through analysis to relate wine characteristics to grape characteristics. We will mention it many times, but always keep records of observations and intermediate results throughout the winemaking process. This will allow you to review historical actions and isolate problems back to the offending step and enable changes to improve the process.

It is important that the harvested grapes are in good condition and free of material other than grapes (MOG). If you do not have sorting tables and optical scanners, hand sorting may be an additional task for your picking crew (Figs. 4.15 and 4.16). Be sure to give them detailed instructions.

Be sure to celebrate the harvest! It will likely include a lengthy crush within the same 24 h. This is a substantial accomplishment after considerable work throughout the season. There were undoubtedly many challenges from the weather or pests or any number of potential threats. A wonderful meal and great wine are a nice way to thank the picking crew and harvest team.

4.3   Step 3: Crush

The harvest day is not done until the grapes make it through the next step of destemming and crushing. Figure 4.17 shows our small destemmer/crusher. It is mobile and easy to clean. Note how the word crush is an overstatement. We just want to squeeze the grapes enough to break the skin. This is accomplished by the rollers

Fig. 4.15 Even after field sorting, take the time to quality sort and screen the grapes before entering the winery

shown in Fig. 4.18. There is an internal auger (Fig. 4.19) that rotates and pushes the stems out the side and the crushed berries fall through the properly sized holes. There are various options to either ferment whole clusters or individual berries depending upon your variety and desired wine style. If you want more fun, a bigger mess, and probably a lower yield, you can crush using your feet or hands. Larger commercial operations may use approaches that destem first and crush second, but even though the sequence is reversed, the same essential functions are performed. Both approaches can produce quality wine.

This operation should be performed as soon as possible after picking the grapes. The equipment should be ready and already be clean from last season's use and a year of storage. Perform an inspection of equipment prior to use in the harvest. The site should be prepared to facilitate a continuous flow with a minimum number of steps from the vineyard to the fermentation tanks. Pre-sort the grapes to minimize leaves and MOG.

Figure 4.20 shows dumping the sorted grapes into the hopper. The hopper's paddle moves the grapes along into the crusher. Whether your operation is big or small, cross-train multiple operators for safety and efficiency.

Use food grade pumps, hoses, and buckets in all juice transfer operations. This juice slurry of sugary pulp and skins is called the must. Transfer the must to tanks

Fig. 4.16 If you cannot afford optical sorters, use multiple teams to help keep up the pace of this hand sorting operation

Fig. 4.17 Mobile destemmer/crusher facilitates setup, use, and cleanup each and every season

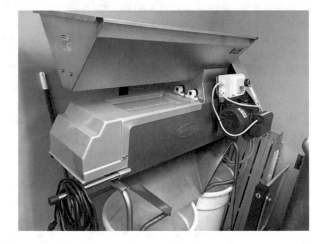

Fig. 4.18 Rollers will crush the grapes by squeezing just enough to break the skins

suitable for fermentation and you are ready for the next step. Our must starts full of flavor and high concentrations of sugar.

The red color in red wine of course comes from the skin. More time means more color. Some choose more time to get more color and other skin effects through a process called extended maceration. If you are making white wine or a Rosé from red grapes, you should reverse steps 4 and 5. This means press right after you crush and before you ferment.

Figure 4.21 shows a list of various crushing destemming tips. Create your own and in time, build a checklist. Once you have crushed your grapes or have decided to go with whole clusters or some combination thereof, it is time for fermentation.

After what was probably a very long and challenging day, be sure to celebrate a successful harvest and crush! Whether you are a big producer or small, you and your crew will enjoy quite an accomplishment as in Fig. 4.22.

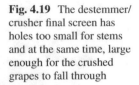

Fig. 4.19 The destemmer/crusher final screen has holes too small for stems and at the same time, large enough for the crushed grapes to fall through

4.4   Step 4: Ferment

Add yeast to grape juice and fermentation will dominant the scene. This is the process of yeast converting sugar to alcohol, CO_2, and heat. Add yeast to the grape juice and the microbial process converts the glucose and fructose sugars present in the wine grapes to ethanol, energy, and other byproducts. Adding commercial yeast in the powder form requires hydration as shown in Fig. 4.23. Let us discuss some of the key drivers to the overall fermentation process that will be the foundation of our various root cause analysis tools.

One of the key drivers is the type of yeast. Yeasts may be found naturally in the vineyard or they may be purchased commercially. Year after year of using natural yeasts enable an understanding of the ultimate wine characteristics. Many great wines undergo fermentation utilizing yeasts found naturally on the grape skins. One of the dominant yeasts is Saccharomyces which has a relatively high tolerance to ethanol, while other yeasts may not.

Winemakers may choose various properties of commercially grown yeasts. These usually come with an advertised temperature range, alcohol tolerance levels, and comments and recommendations for use with certain varietals, flavor, and body

Fig. 4.20 Ease of operation enables good flow

Fig. 4.21 Crushing and destemming tips

Crushing Destemming and Tips:

1. Include a final sorting at the crush pad before the grapes enter the hopper. Note that sharp eyes and subjectivity can vary from person to person so include multiple checkers.
2. Don't overload hopper for smoother continuous operations
3. Compost damaged fruit and rejected stems and leaves.

weight characteristics. These also come with instructions on the procedures for inoculating the wine, adding yeast food, and providing the best environment for yeast growth. Different yeasts have different rates of sugar conversion and reactions depending on grape juice constituents. (Yeast selection criteria are discussed in Sect. 5.4.2). It is important to add yeast food approximately half way through the process of reducing brix from initial to near zero. If using commercial yeast, prepare the yeast and yeast food for hydration and inoculation per the manufacturer's recommendation.

Fig. 4.22 Take a few moments to recognize the team; smile and reflect on your accomplishments!

Fig. 4.23 Hydrate the yeast as a part of inoculation

Juice temperature is another key driver of the fermentation process. Note that the conversion of grape sugars to ethanol by yeast under anaerobic conditions is an exothermic process which means that it releases heat. Heat causes the yeast to convert the sugar to alcohol and carbon dioxide faster. Fermentation rates increase exponentially with temperature. Above 104 °F, the yeast will not survive and the fermentation will stop. It is necessary to control the fermentation within an optimal range. A fast fermentation will result in less time for tannin, color, and flavor extraction from the skins, so strive to keep the temperatures down for optimum extractions. Faster or slower fermentation rates is a tool the winemaker can use according to the style of wine you want to make.

Fig. 4.24 Fermentation
tips

Fermentation Tips:

1. For red wines or chardonnay, consider employing malolactic fermentation, a secondary fermentation to convert tart tasting malic to lactic acid. Can be initiated with the addition of a lactic acid bacteria.
2. Punch down 3 to 4 times per day or provide a flow over path for juice over the cap to prevent drying and enhancing color extraction
3. Keep a log of temperatures to not only act for hot extremes that could kill the yeast but also to track trends to help plan and prepare for upcoming activities

The fermentation rate is also impacted by the initial concentration of sugar and pH of the juice. Temperature sensitivities are also impacted with varying pH levels. As the pH decreases corresponding to more acidic condition, the yeast organisms operate more efficiently and convert faster.

Fermentation tips are presented in Fig. 4.24.

No sugar or very low levels of residual sugar is the new characteristic of our grape juice. The sugar has been converted to alcohol in a must solution. We need to move onto an important process to clean up this must and separate out the wine from the unwanted skins. We are now ready to press.

4.5   Step 5: Press

This is the process of pressing or squeezing the fermented wine and disposing of the skins and seeds. Typical presses are either wooden, mechanical screw presses or stainless hydraulic bladder presses. We are engaged in a dry red winemaking process so after many days, all of the sugar will have been fermented to alcohol and we are ready to begin the press. The timing and decision on when to press is another critical event in winemaking. Not only will the sugar have been converted, but the color making maceration process of mixing the phenolic from the skins with the pulp and juice from the berries will come to an end.

More time on the skins produces more intense colors and tannins. We once used a relatively fast working yeast on a Pinot Noir vintage and decided to press as soon

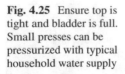

Fig. 4.25 Ensure top is tight and bladder is full. Small presses can be pressurized with typical household water supply

as the sugar was gone after about 3 days. That particular vintage has little color and body as compared to our other wines with maceration periods between 6 and 8 days.

To avoid oxidation of the finished wine, it is best to press as soon as the wine is finished fermenting. If you want extra time on the skins, do a cold soak before inoculating with yeast instead of extending maceration after fermentation.

As shown in Fig. 4.25, we use a bladder press which provides nice continuous pressure. The picture in Fig. 4.26 shows the interior black bladder or balloon that is filled with water and expands radially outward and pushes the grapes against the porous outer metal circular housing. The juice or wine flows through the sides, drips to the pan, and flows out the port into the transfer bucket or plumbing. Pressing tips are summarized in Fig. 4.27.

Prior to the bladder applying pressure, the first wine flowing out is referred to as free run. Some winemakers may keep this separate from the pressed juice with the belief that this may taste less bitter due to less impact from the hard-pressed skins. In our operation, we combine both together. Sampling our day 1 wine showed a lot of fruity but also vegetative aromas.

Pressing is that strange balance of firmly but gently squeezing the grapes to extract our newly fermented alcoholic wine. This new day of our winemaking journey readies the wine to begin the complex process of aging.

Fig. 4.26 Hydraulic press
provides uniform press
action

Fig. 4.27 Press tips

Press Tips:

1. When using a bladder press,
    make sure the actual action
    area or grape surfaces that
    squeezed are uniformly
    loaded around the perimeter
    and elevations of the
    equipment to help
    maximize yield.
2. Taste the wine at many
    stages throughout the
    overall process and keep a
    mental or written record to
    track vintage to vintage
    quality.

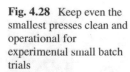

Fig. 4.28 Keep even the smallest presses clean and operational for experimental small batch trials

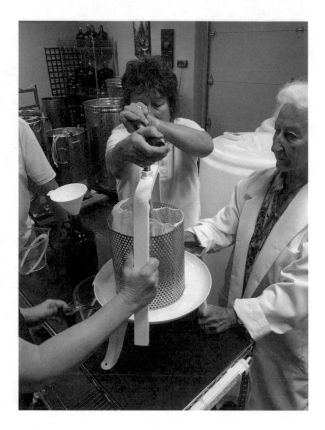

A very small press, like the one pictured in Fig. 4.28, is ideal for processing small batch trials. As depicted in Fig. 4.29, press in an area where you can wash down equipment and the floor easily. And it is always good to have another set of eyes to independently check on quality as shown in Fig. 4.30.

4.6   Step 6: Age

We want the wine to be stable before it is bottled. While we are stabilizing, we also want to enhance some stylistic flavors or aromas. Aging refers to the post-fermentation and post-pressing process that provides time for chemical reactions to occur and improve the stability of the wine. This can be done in a variety of containers ranging from oak barrels to stainless tanks depending upon the style and flavors you want to impart. Wooden barrels or containers for storage of wine dates back thousands of years. We use oak barrels not just to store and transport wine, but to actually enhance the flavors and stability.

Oak contains many chemical compounds that can add flavor and personality to the wine. These can include vanilla, toast, and tobacco notes. The wine can become

Fig. 4.29 A convenient crush pad close to the winery facilitates press operations and cleanup

more mellow and complex after months in contact with oak barrels. We do use French oak barrels, however, the science does seem to indicate that for a particular oak species, whether it comes from an oak tree in France or Hungarian, or America, will impart the same flavors and aromas. As shown in Fig. 4.31, the age of the barrel will dictate how much oak extraction occurs with time. The construction of the barrel or cooperage is also critical. Sediments or other particulate products that might be formed during aging can be removed by racking, filtering, or fining clarification processes. Racking refers to the process of pumping out the wine into a clean container and removing the sediment and cleaning the aging container. Gravity forces the particulates and sediment to the bottom of the barrel.

Stainless steel tanks, like the one depicted in Fig. 4.32, are ideal for unoaked wine varietals or to temporarily transfer or blend wine. You can add oak chips or staves to impart the flavor characteristics that oak provides, but since there is no micro-oxygenation occurring through the wood barrel, some of the subtle differences will not be achieved.

Aging should occur with little to no oxygen. This helps polymerization and stabilization of color and softening of tannins. In many containers like oak barrels, there is a moderate loss of wine through evaporation through the semi-porous wood. Over the course of 18–24 months of aging in our 60-gallon oak barrels, we have found that we lose almost 3–4 gallons of wine. This "angel's share" needs to be replenished or topped off to prevent excessive exposure to oxygen. We do this by

Fig. 4.30 Mom and dad keep a close eye on quality and have no problem with operation critiques

adding wine from carboy glass containers containing the same wine and stored separately.

Heavy and unwieldy carboys, like those shown in Fig. 4.33, should be positioned in locations with easy access. The stainless tanks have convenient adjustable lids which allows us to age different volumes of wine each year. We also use these stainless tanks in the first few months to get through the first or second racking. For short periods of time, when we have run short of top-off wine, we have added inert gas such as Argon or Nitrogen to displace the oxygen and limit oxidation.

We generally try to age our Bordeaux varietals at least 18 months in oak and reuse barrels to get through 2 or 3 seasons. The used barrels impart much less flavor each year. By the 3rd or 4th year, the barrels can be considered neutral in terms of imparting oak flavors. Aging reactions are affected by temperature. We try to age at cool to moderate temperatures, between 10 and 20 °C. An increase in temperature in general increases the reaction rate. If a reactant is volatile and we heated the wine, the concentration would drop. Some product formation may be inhibited with colder temperatures. Moderate temperatures generally provide a balance between these two effects.

Fig. 4.31 Note that new oak barrels bring nice but heavy oak vanilla flavors. Barrel aging also requires monthly or more frequent maintenance

Aging for 18–24 months may not seem long in the life of a fine wine. However, the costs of proper storage are adding up and it may be ready to bottle and get it out the door. Taste the wine and rely on your own judgement, and if you are fortunate to have a track record of wines, refer back to the historical quality of wine to determine readiness to bottle. Before we bottle our young wine, we want to ensure stability and move onto the clarification step.

Age in oak long enough to get the oak flavors you like; refer to Fig. 4.34 for tips. Additional aging in the bottle may be necessary to soften the tannins and complete the aging duration.

4.7   Step 7: Clarify

Our wine may still be young, but it has matured well through the early aging process. We are now ready for the clarification step. There are a variety of options that fall into this clarification step. These include racking, filtering, fining, centrifugation, refrigeration, pasteurization, or other assorted methods. In general, clarification is the process to remove insoluble elements from the wine before bottling. In

Fig. 4.32 Stainless barrels
with adjustable lids
provide convenient
accommodation of
different volumes

our winemaking, we have typically used racking and filtering. We introduced rack-ing in step 6 to make a note that it should occur early in the aging process or soon after fermentation is complete. This is to get the wine off the gross lees or at least most of the dead yeast. Unwanted insoluble elements may include dead yeast, bac-teria, titrates, and other miscellaneous debris.

The wine stability will improve by the removal of these interlopers. Wine that is allowed to age on these lees may develop uncontrolled off-tastes and aromas. There is a trade between complex flavor infusion with longer time on the lees versus insta-bility due to longer contact with disruptive compounds.

As discussed earlier, racking is moving the wine from one container to another to allow the removal of the residual sediment at the bottom of the barrel. This can be done by gravity or transfer pumps.

Some prefer "gravity only" for minimal impact to the wine; however, we have used food grade pumps, plastic hoses, and other siphoning devices for better effi-ciency and convenience and we do not expect any discernable degradation in the quality of the wine. These filters can remove large particles that impact clarity and or spoilage organisms that impact stability. This equipment allows the use of differ-ent size filters and from one to many. The key is finding the combination that pro-vides a clear bright wine with complex flavor profiles but capturing all of the bacteria that could cause spoilage.

Fig. 4.33 Carboys ready
with wine to top-off barrels

Fig. 4.34 Aging tips

Aging Tips:

1. Rack the wine during the first month
   after fermentation to get the major
   sediment and dead yeast out of the
   wine
2. Monitor wine frequently to ensure
   barrels are topped off to avoid oxygen
   exposure at least monthly and
   preferably check weekly.
3. Keep the aging tanks or barrels in a
   cool temperature-controlled
   environment to avoid heating and
   damage to the wine. A constant
   unvarying temperature is also
   preferred to limit material expansion
   and contraction causing container
   leaks.

We have used a sub-micron filters and pump system shown in Fig. 4.35. This
employs a pump to force the wine to flow through a filter membrane with approximately 0.45-μm pores. This was once considered the appropriate size needed to
strain out even the smallest bacteria and thus accomplish sterilization. However,
recent discovery of even smaller bacteria has shifted this down to approximate 0.2-
μm membrane filters.

There may be a concern that a filtration process removes flavor and aroma compounds. However, many expert tasters have not been able to distinguish between
filtered and unfiltered wine.

Fig. 4.35 We use a porous membrane filter just prior to filling bottle

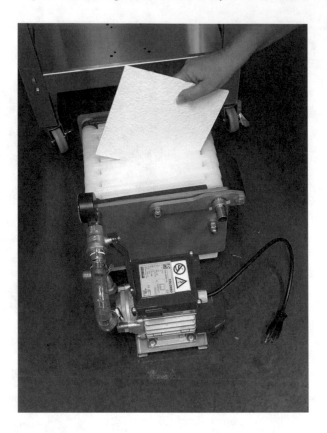

Fining is another clarification process that allows the winemaker to target unwanted elements that might be in the wine in a soluble form. This is the process of adding an adsorptive agent and then follow-up with removing the precipitation of the additive by racking or filtration. The unwanted elements bind with the agent and then settle out. It is possible to remove haze forming elements like proteins or tannins to soften the wine or metal ions. The winemaker needs to thoroughly understand the chemistry of this process to not introduce adverse treatment of the wine.

Our winemaking clarification tips shown in Fig. 4.36 help to stabilize the wine, removed any cloudiness, and unwanted bacteria. Our yummy wine should not just be stable, but still very delicious and ready for bottling.

4.8 Step 8: Bottle

Bottling is our next important step. The goal is to efficiently fill the bottle and install the closure for reliable short- or long-term storage. Whether for small or large operations, we want to do this quickly to minimize excessive air exposure as well as

Fig. 4.36 Clarification tips

Clarification Tips:

1. It is important to have conducted a robust racking program starting post - fermentation and maintained throughout the entire barrel or tank aging process.
2. Clean and sanitize all components that come in contact with the wine leading up to final fill of bottles.
3. If you have a history of unstable bottled wine or many bottles coming back, consider a more aggressive clarification program that includes a variety of methods. This means in addition to racking, consider filtering and fining or another combination

Fig. 4.36 Clarification tips

minimize costs. Large operations may have their own bottling assembly lines or may hire an outside bottling truck to drive in and bottle on-site.

Organize the facility, tools, and materials prior to the start of operations. Operations in close proximity minimize time and motion between stations and convenience of workers. Clean and sanitize the bottles with a potassium metabisulfite solution before use.

The equipment should include an efficient fill of the bottles directly from the tanks and barrels. Our transfer hose draws wine directly from the oak barrel or blending tank and fills a staging tank in the top of the bottler as seen in Fig. 4.37. This tank has an electric float with minimum and maximum sensors to automatically refill when the volume gets low.

Gravity fed lines shown in Fig. 4.38 fill each of four bottles with individual plumbing and individual automatic shutoff valve assemblies. This ensures the same correct fill level of each bottle.

Despite this automation, we do a quality check to ensure that each bottle is filled to the correct level. This is roughly a 1/2 in. air gap between the base of the installed cork and the top of the wine in the finished bottle. This, of course, is best checked before installation of the cork. The inspector shown in Fig. 4.39 is visually inspecting the high point of the wine in the bottle against a standard bottle already set at the correct height. This inspection can be a sample of bottles going through the line and adjustments made as required.

For our early small volume and production rate, the single bottler was the slowest operation in the line. After we grew to filling four bottles in parallel, the simple operation of installing the cork in the bottle was the slow point in the overall bottling assembly line. We then expanded to two parallel manual corking machines, as shown in Fig. 4.40. Establish you own production rate goals and assess the efficiency and

Fig. 4.37 Our four-bottler has a staging tank at the top which refills via an electric float

duration of each your production steps. Efficient operations need processing times through each station to be compatible with the overall production line rate.

We then apply a heat sensitive shrink wrap covering to seal the bottle and cork. This step is shown in Fig. 4.41. Wax can also be used, but it is more of a mess when applying as well as when removing it to drink.

The next step, as shown in Fig. 4.42, is to apply the labels which we have printed and made available before we start the bottling line. We construct a digital image for each particular wine type with the appropriate label data. Labels are printed on label paper with correctly sized peel off sections and adhesive on the back to allow manual application. The labelers are peeling off each label from the pre-printed stock, placing, and orienting the label squarely on each bottle. We use back labels on our red blend to describe the blending ratios for each particular vintage even though it requires twice the printing and installation labor. If you do not have automation or a jig for accurate alignment, you will need to rely on people with a sharp eye and patience to get these two labels straight. Refer to Chap. 2 or your local wine laws for the required information on labels. Finally the bottles need to be loaded back into the empty boxes that have made their way from the front of the bottling line. Be sure to move the finished cases into your temperature-controlled storage as soon as possible.

It is also important to have a production line that has adjacent stations to allow the product to move seamlessly from one stage of assembly to the next. It may be

Fig. 4.38 Improve the overall flow of the complete bottling line by assessing each stage limit. Equipment upgrades, well trained, and happy operators helped us increase bottling rate fourfold

Fig. 4.39 A bottling line includes a periodic sampling check of the fill level to verify the wine height in the bottle to provide the proper air gap in the finished corked bottle

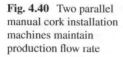

Fig. 4.40 Two parallel manual cork installation machines maintain production flow rate

linear or u-shaped or whatever fits naturally into your own facility. The key is to minimize down time moving from one station to the next.

Figure 4.43 summarizes our bottling tips. Our wonderful wine is now well protected and ready for safe keeping until that future special moment when it will be uncorked and ready to drink.

4.9   Step 9: Store

We are on the home stretch. There is still a critical step remaining for this wine which has just begun its life in the bottle. Yes, we have made this beautiful wine and it is either ready to drink or ready to be laid down for some additional aging time in the bottle. We want to take precautions to make sure this is just the right

Fig. 4.41 Careful application of shrink wrap covers to encase the bottle top

environment. Our goals are not too hot, moderate humidity, no vibrations, minimal light and of course, easily retrievable for that magical moment when it is ready to drink.

There are many kinds of bottle storage places but perhaps only a limited few in your current conditions. Our case storage is on a cool tile floor next to the concrete wall of our house foundation. It has a very stable temperature range, even in the heat of summer.

Individual bottles are stored in a temperature and humidity controlled cellar (Fig. 4.44). The air-conditioning unit vents to the outside. Not all of the above goals may be accomplished, however, the more you can check off, the better life you offer your valuable wine. Depending on your location and climate, you must assess when and if you might need cooling during the summer months.

Pay attention to the orientation of the bottles. Corks down or sideways is preferred so that the cork does not dry out and leak over many years on the shelf. Garages and warehouses may have the room, however, they often have little insulation which result in temperature swings and in many regions, may run too hot during the summer months. Even if you only have a few bottles in your collection, storage on top of the refrigerator will likely be too hot.

Refer to Fig. 4.45 for our storage tips.

Congratulations on completing the review of basic quality grape growing and winemaking. A finished quality glass of wine is always a joy to share with family and friends (Fig. 4.46). At the same time, we also recognize it is not all smiles and

Fig. 4.42 Our last bottling line station is application of the labels. The labelers confirm they have the correct label for the wine coming through and place and orient the label squarely on each bottle

laughter. Farming and making wine can be a daily grind, back breaking, and occasionally heartbreaking. To borrow from a 70s rock group, there can be a little blood, a lot of sweat, and a few tears. However, we firmly believe you can still have fun and enjoy the results. There is a wondrous pleasure in figuring out the root cause and contributing factors to the many problems you will face. The quality wine you make will bring happy memories and a lasting smile.

Fig. 4.43 Bottling tips

Bottling Tips:

1. Clean equipment and sanitize bottles before use
2. Periodically check fill level of bottle and adjust if low or high
3. Keep the crew motivated with a nice wine and meal after the bottling session is complete

Fig. 4.44 Environmentally controlled wine room with temperature below 59 °F and relative humidity above 60%

Fig. 4.45 Storage tips

Bottle Storage Tips:

1. Immediately after bottling, keep bottles vertical with corks up for a couple of days to allow bottle to reach equilibrium and then turn sideways for longer term storage. Spot check for leaks.
2. Clearly label boxes before stacking and storage for quick look inspection of what's in each box
3. Wine cellar software to track bottles is beneficial. An early first step is to organize location of bottles by a simple logic you can remember. There are many possibilities. Alphabetical by varietal or blend, chronological by vintage, geographic region of vineyards and or producers, or an easy combination of some of these. Pick an approach that is easy for you to remember.

Fig. 4.46 A quality glass of wine always brings a nice smile

Chapter 5
Red Wine Volatile Acidity Problem Solving with Cause and Effect Analysis

This chapter tackles high volatile acidity (VA) as the effect for which we will apply RCA to determine the root cause. We start with VA basics, discuss our VA scenario problem statement, and apply the cause and effect diagram. We identify possible causes, and assign likelihoods. We will point out key tests and discuss preventive measures. Finally, we will organize and prioritize the possible causes. This process will first widen the number of possibilities and then narrow the field. We will work toward identifying contributing factors and a true root cause.

We sort these candidate drivers in priority order to focus our time and resources on resolving the leading candidates. We do not want to waste time on unlikely causes. With practice, we will make better and better wine.

5.1 Red Wine Volatile Acidity Basics

Total acidity is the combination of fixed and volatile acids. Fixed acids include malic, lactic, tartaric, and others. Volatile acids are gaseous and are most often associated with spoilage. One of the most common is acetic acid. Acetic acid results in a vinegar taint or fault. It can originate in the field or be produced or enhanced during fermentation or aging. It can come from spoilage yeasts or bacteria that are also exposed to air. There are various spoilage microbes, but Acetobacter is one of the most common. Without proper training, it is easy to miss the presence of unwanted spoilage causing pests and diseases. In Figure 5.1 are Pinot Noir grapes appearing reasonably healthy. Some signs of powdery mildew but various other natural yeasts and microbes are not visible to the naked eye. Use your quality winemaking skills to showcase the vineyard and its terroir, as opposed to allowing or introducing spoilage conditions to thrive and degrade your wine's quality.

When you do not want vinegar, you must protect against the microbes, oxygen, and high temperature environment in which they thrive. Too high VA levels is bad

J. Steakley, B. Steakley, *A Quest for Quality Wine, Every Time*,
https://doi.org/10.1007/978-3-030-34000-1_5

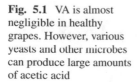

Fig. 5.1 VA is almost negligible in healthy grapes. However, various yeasts and other microbes can produce large amounts of acetic acid

but a low level might be ok. The legal limit for VA in the USA is 1.2 g/L for reds. Understanding and prevention of excessive VA is our goal. If your goal is to make a lovely red wine vinegar, then go for it. If your goal is to make a beautiful well-balanced wine with low VA, then pay attention.

5.2   High VA Problem Statement and Cause and Effect Diagram

Now let us capture the effects and define a clear problem statement as shown in Fig. 5.2. We might begin with a problem statement that reads like this: "The wine smells like vinegar but still tastes pretty good." This is an okay start because it has a comment on the taste as well as on how it smells, however it is limited. We still can enhance the description and timing specifics.

Provide detail regarding the grape varietal, the source of the grapes, the vintage of the grapes, and the phase of the processing for when the problem was first observed. More details can and should be added as the investigation matures. The next edit might read like this: "This Cabernet Sauvignon wine was made from 2013 grapes from the Steakley vineyard and smells like vinegar just before bottling." As the investigation progresses, add more details. Attention to details will pay off.

Fig. 5.2 Our VA scenario
starts with an informative
problem statement
describing conditions and
effects

- This wine smells bad like vinegar
- This wine still tastes pretty good.
- Smelled like vinegar just before bottling
- Harvested in 2013
- Cabernet Sauvignon
- Source from Steakley Vineyards, Santa Cruz mountains
- Sugar at harvest 26° Brix
- Titratable Acid level 7.8 g/L
- pH 3.49
- Volatile Acidity 0.1 g/L

Details can include measurement data, vintage, varietal, grape source, grape conditions at harvest, timing of fault observation, and more.

Problem statements should be informative by including descriptive information about grape origin, wine chemistry, and production history. It is important to include both qualitative sensory observations and quantitative data. This allows tracking with respect to desired style as well as wine chemistry trends. Compare with published standard ranges as well as changing values with respect to your own wine history. Highlighting changes, trends, and inflection points and relating to production events can provide insight into finding the true root cause.

Figure 5.3 is the winemaking cause and effect fishbone diagram we will use to organize, track, and solve our high VA problem. The problem statement is placed at the head of the "fish."

5.3 Process and Red Wine VA

This subsection focuses on the process phases, the first of six major bones on the fishbone diagram. These could also be called actions or methods. These are the step-by-step methods to make the wine and can be depicted as sub-bones off the main process bone. The process category is not equipment or materials, but it is the actions or activities in winemaking. For example, this might include the action of barrel cleaning, but not the different barrel materials themselves. Or it might include the action of hydrating the yeast before fermentation, but not the various yeast material type. Figure 5.4 shows the nine process phases with the first seven still on as possible origins of our VA problem and the last two crossed off because the problem was observed before they even occurred.

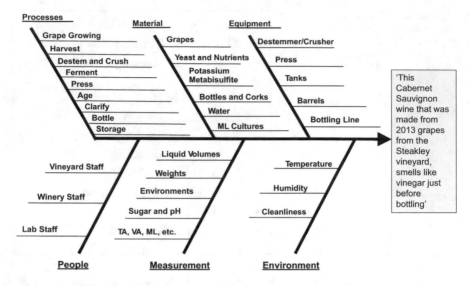

Fig. 5.3 Establish problem statement in the generic fishbone diagram

Fig. 5.4 Knowing which phase the fault was observed, allowed us to drop those that followed

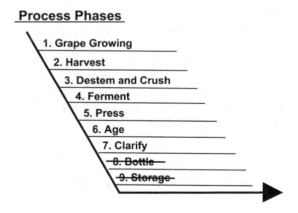

Those detailed step-by-step processes can lead to variation and issues, so they will need to be evaluated in more detail. We have already defined nine major process phases on the generic fishbone diagram. The first step would be to add "sub-bones" or specific sub-processes to the fishbone diagram as shown in Fig. 5.5. For example, the sub-processes of "grape growing" would include winter pruning, watering and fertilizing, spraying and other pest management practices, canopy management, and testing the grapes for harvest. It would be OK to break them down even farther as we go through the RCA process if it helps to find the true root cause. Remember, the devil is in the details.

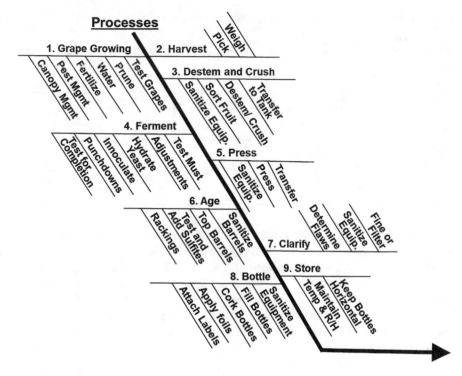

Fig. 5.5 Sub-processes are added to the nine major processes

These sub-processes should all be documented so that every employee assigned to perform them will do so with the same results. This also means that the processes need to be "proven" effective, e.g., if you follow this process for barrel cleaning and sanitizing, we know it will be effective because we have taken microbial swabs and had successful test results.

This scenario detected high volatile acidity before bottling. This fault observation timing allows us to narrow the field and drop the last two phases of bottling and storage. If we did not notice the high VA until opening a bottle after years in the wine rack, we would of course have to investigate time in the bottle and storage.

Even though we can drop the last two phases of bottling and storage, we cannot necessarily drop all the early phases. Since the minimum threshold for smelling vinegar in this scenario will differ person to person, we need to gather more clues and evidence to narrow our focus. One clue would be to look back through lab data for volatile acidity measurements. Table 5.1 lists four hypothetical examples, demonstrating in Example 1 that a significant load of spoilage microbes was introduced during the clarifying process, vs. Example 2 where VA shot up during the aging process. Example 3 points to the fermentation phase as the trigger point for high VA and Example 4 is clearly a microbe source in the vineyard.

If you have extra evidence of this nature you can zero in on all the processes performed during that phase to look for sources of microbes and excess oxygen.

Table 5.1 Examples of different volatile acidity measurement scenarios

Volatile acidity data review (g/L)				
Process phase	Example #1	Example #2	Example #3	Example #4
Grow	0.0	0.1	0.1	0.6
Harvest	0.1	0.1	0.1	0.6
Crush	0.1	0.1	0.1	0.7
Ferment	0.1	0.1	0.7	0.8
Press	0.1	0.1	0.7	0.9
Age	0.1	0.9	0.8	1.3
Clarify	0.1	1.0	0.9	1.4
Bottle	0.9	1.1	1.0	1.4
Store	1.0	1.2	1.1	1.6

And you can skip over earlier phases if the evidence is clear that they are not the source of high VA and therefore not the root cause.

The same rationale can be applied as you move through the other five major bones of the fishbone diagram—assess all the applicable materials, measurements, people, equipment, and environments for the relevant time phases. If you do not have definitive proof to rule out a time phase, then you will have to evaluate it. For purposes of illustration, we will proceed through evaluating all nine bones on the processes major bone.

5.3.1   Process Phases 1 and 2, Grape Growing, and the Harvest

Growing the grapes and the picking date determine ripeness. The foundation of wine aromas and flavors originate from the grapes. The natural differences between grape varietals provide for the complex flavors and aroma profiles that we enjoy. These flavors and aromas are what we strive for. However, it may be unwanted fungus or microbes on the skins that give us problems. Given long growing seasons, exposure to a diversity of environments and pests, this can be a risky phase for VA. Good news! There are a number of preventive measures shown in Table 5.2 that can make these unlikely.

Each of these possible causes can be controlled and reduced to low likelihoods after application of various preventive measures. Effective spraying can control pests, fungus, and microbes. At the same time, it is important to recognize all protocols are not 100% effective. The details of your spraying might not cover all vine leaves and shoots. SO_2 concentrations may become ineffective near wine surface and even low-level oxygen headspace layers in aging barrels. In our scenario we are also assuming the grapes got sufficient sun and water to ripen. Apply multiple methods and follow-up by checking for results and adjusting as necessary.

Apply tests to aid in detection. Detect early with frequent walks and inspections. Really get to know the living ecosystem of your vineyard. This can occur during

Table 5.2 VA grape growing and harvest causes can be reduced to unlikely with preventive measures (PM)

Possible causes	Likelihood after PM	Comment
Pests	Low	Most are not direct source but contribute as vectors that spread infestation
Fungus	Low	Moderate but low after controlled organic spraying
Natural Microbes	Low	Moderate but low if application of SO_2 addition to must
Natural acid levels in grapes	Low	Low for direct VA producer. Assuming application of correct fertilizer
Insufficient nutrients	Low	Low after application of water and fertilizer to soil

Table 5.3 Useful tests to aid in detecting possible VA causes in grape growing

Possible causes	Tests	Comment
Pests	Visual inspection	Pests may transmit disease, damage leaves, and grapes. Frequently walk the vineyard. Inspect the vines. Compare with published pest descriptions. Check neighbor experiences
Fungus	Visual inspection and laboratory tests	Fungus growth may indicate high bacterial growth conditions and spoil grapes. Frequently walk the vineyard. Inspect the vines for symptoms. Compare with published fungus descriptions. Check neighbor experiences. Submit samples for laboratory
Natural Microbes	Laboratory tests	Microbes may produce unwanted acetic acids. Frequently walk the vineyard. Inspect the vines for symptoms. Submit samples to laboratory. Check neighbor experiences. Moderate but low after application of SO_2 addition to must
Natural acid levels in grapes	Acid types and levels	Understand baseline of acids. Test levels and types of acids in grapes at harvest
Insufficient nutrients	Vine inspection and petiole analysis	Insufficient nutrients may impact growth and produce unwanted products. Inspection of vine, leaf, and grapes. May cause yellowing in older leaves and/or stunted growth. Reference website for visual signs. Petiole Analysis for analytical assessment

winter pruning, before and after bud break, during canopy management, wire training, preparing for harvest, and securing for the next season. This should cover the entire growing season.

There are various good online sources for detailed reference material on pest control (University of California Agriculture and Natural Resources 2019). One of the keys is early detection and application of preventive measures, which leads us to Table 5.3. Size the application to best fit the type and level of your particular infestation.

We are striving for an effective, practical, and affordable implementation. With preventive measures in place within the vineyard, we will be harvesting healthy grapes. We do not want excessive fungus nor bacteria on the skins. Sorting and discarding broken or moldy grapes in the vineyard is a useful quality step as shown in Fig. 5.6. We also recognize that in our real-world vineyard, all processes and preventive measures are not 100% successful.

Table 5.4 lists various possible high VA causes and provides a summary of preventive measures. In addition to good vineyard practice, it is important to add SO_2 after the crush. Our action is to make sure the preventive measures are executed in a thorough and timely manner in the vineyard. Our assumption for this scenario is that we will have walked the vineyard near daily, implemented reasonable vineyard practice, tested, and conducted good harvest practice.

In summary, Phases 1 and 2 are unlikely causes and assigned a low likelihood. Additional comments on likelihoods for these first two process phases are listed in Table 5.5. Without application of some or most of these preventive measures, your investigation may find a higher likelihood contributing factor that makes it on the prioritized list of high and moderate candidates.

5.3.2   Process Phases 3, 5, and 7: Destem and Crush, Press, and Clarification.

Phases 3, 5, and 7 are dominated by the use of equipment and all consist of relatively short wine contact. They do not deliberately add materials to the wine. The press process phase and the clarification phase will remove materials; however, these actions should not be a direct increase in the level of VA. For the scope of our VA scenario, we are not covering nor considering more invasive fining operations since they are needed to correct specific flaws other than VA. Clarification is being accomplished by filtering through porous mechanical filters.

Fig. 5.6 A careful inspection or pre-sorting for moldy or damaged grapes in the vineyard is simple and helpful

Table 5.4 Preventive measures for VA in grape growing and harvest phase

Possible causes	Preventive measures	Comment
Pests	Beneficial insects to control adverse insects (ladybugs to control aphids which are vectors of disease)	Size the type and number of insects to service the type and number of pests
Fungus	Organic sprays. Stylet mineral oil is effective	Assess the level of infestation and size the concentration, volume and frequency of spraying
Natural microbes	After crush, add standard SO_2 (by adding potassium metabisulfite) at 50 ppm	Measure free SO_2, pH, and determine addition to get to desired level for your condition
Natural acid levels in grapes	N/A	Test levels and types of acids in grapes at harvest to track changes thru winemaking
Insufficient nutrients	Fertilize	Petiole analysis to determine nutrient status enabling type and quantity of fertilizer

Table 5.5 Likelihoods for this VA scenario 1, process phases 1 and 2. All are low likelihoods for this scenario in which we applied these preventive measures

Possible causes	Likelihood for scenario 1	Comment
Fungus	Low	No visible signs after spraying with organic mineral oil
Natural microbes	Low	Used proper sulfur dioxide levels after fermentation
Insufficient nutrients	Low	Watered and fertilized regularly
Pests	Low	Controlled vineyard
Natural acid levels in grapes	Low	Strived for high sugar and alcohol which left borderline low acid levels, but still ok

We have grouped these together because these phases have a common process of sanitizing the equipment before and after each use. The process for cleaning and sanitizing must be well defined and always performed in the same manner. We are also assuming harvesting is done by hand with minimal contact with grapes and clean gloves and cutters. If significant mechanical harvesting equipment is used, we recommend cleaning.

Refer to the winery cleaning and sanitization plan in Table 5.26. Our cleaning strategy is effective, affordable, and covers all phases of the winemaking process. After incorporation of these cleaning methods, the likelihood of these phases contributing to the problem is all low. For this VA scenario 1, process phases 3, 5, and 7 are all unlikely causes given our cleaning and sanitation preventive measures.

5.3.3   Process Phase 4: Fermentation and VA

Wine is fermented grape juice so we certainly must ferment; however, this action may also introduce materials and conditions that can cause excessive VA.

The fermentation process starts with grape juice and glucose and uses yeast to produce alcohol and wine. Figure 5.7 illustrates incoming grapes, along with bacteria and fungus that might be on the skins. The yeast is converting the sugars to ethanol and CO_2 at rates and in ways that depend upon the yeast type, the available food, and fermentation temperature. These processes can be unknowingly altered and produce unwanted outputs in various ways. Let us look at some possibilities.

Natural microbes and natural yeasts will likely come in with the grape harvest. More will be collected in the winery. During the process of fermentation, the sugar is converted to alcohol and the juice becomes wine and then develops a risk of spoilage with exposure to air and oxygen. If we allow the fermentation to begin without yeast food, some of the yeast byproducts will include acetate and create volatile acids. The process phase will use fermentation tanks which are at risk for contamination. Perceptive tests will guide proper application of preventive measures and reduce them to lower likelihoods. You may consider preventive measures even when more formal tests are too expensive and inconvenient. Use your senses. You are a good tester, but like many things, it takes practice. Some spoilage effects produce cloudiness and color change indicators. Unwanted components might produce haziness and more extreme cases could cause filters to clog up or sediment to increase. Changes in color can indicate unwanted microbes at work. It is likely a strong color change will be accompanied by an aroma change, but there are still the rare possibilities of off-color but no off aromas.

There are many smell indicators. Some spoilage effects also produce aromas or gases indicating something is amiss. This might be a result of the bad spoilage organisms. This also might be from good organisms but there are also too many bad

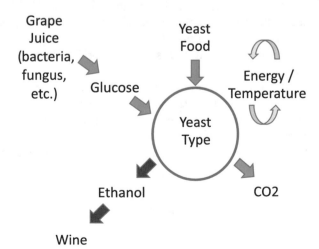

Fig. 5.7 The fermentation process

organisms in the juice or wine. Fermentation and yeasts can produce rotten egg smells (Hydrogen Sulfide). Ethyl acetate smells like nail polish remover. Brettanomyces yeast spoilage can produce mousiness aromas and taste. Lactic acid bacteria metabolic spoilage can produce the odor of geranium leaves. There are many publications and studies regarding wine faults and symptoms which we do not plan to summarize details here, but rather provide an effective strategy, tools, and supporting examples. As a sensitive winemaker, a hint of an off smell also "screams" for further investigation. Table 5.6 summarizes possible causes of VA and comments on the likelihood of occurrence. This commentary is based upon this scenario's assumption of applying preventive measures. Of course, if you do not apply these measures, the likelihood of a particular cause of VA in your wine goes up.

The punch down is one of the key steps to keep the cap of skins wet during the fermentation process. For a small-scale operation for a small batch of wine, it involves pushing the cap down multiple times a day as shown in Fig. 5.8. This will keep the cap wet and most of the unwanted oxygen out. This is an example of a necessary winemaking operation and is an important step to prevent oxidation and conditions promoting VA.

Tasting is perhaps the foremost sense to enjoy wine and at the same time also provides informative clues regarding faults. When drinking wine, tasting and smelling are experienced together. It is true that the number of different tastes or flavors might have an infinite number of combinations, but it is generally accepted that there are five types of flavors: sweet, sour, salty, bitter, and umami. However, this scenario is focused on volatile acidity with an emphasis on volatile or aromatic; we are going to rely more on our sense of smell.

Fig. 5.8 Whether you punch down or pump over, there are many options for keeping air away from wine during fermentation

Table 5.6 The likelihood of fermentation elements causing Volatile Acidity (VA) changes from moderate to low after application of preventive measures

Possible causes	Likelihood	Comment
Bacteria or mold	Moderate but low after PM	Bacteria and molds add risk of VA. Low after preventive measures. Apply good sanitization and use of SO_2
Yeast inoculation	Moderate but low after PM	The fermentation process needs to add yeast for healthy process. Prepare culture per specifications. Control temperatures.
Nutrient deficiency	Moderate but low after PM	Low nutrients can result in stuck or unhealthy process and produce excessive acetic acid and VA.
Air	Moderate but low after PM	Growth rate of bacteria increased with oxygen. Keep cap wet and the air out.
Dirty equipment	Moderate but low after PM	Fungus or bacteria may contaminate wine and equipment if not sanitized and cleaned. Sterilization is not required for winemaking, but still need a good sanitization level of cleanliness

Table 5.7 Tests that aid in determining VA cause and contributing factors from fermentation

Possible causes	Tests	Comment
Bacteria or mold	During fermentation: Smell Brix vs. time Temperature vs. time Specific gravity vs. time Post-fermentation: Sensory: Color off Sensory: Aromas off Sensory: Tastes off Microscope images	Want healthy conditions Off colors, aromas, and tastes can be produced by microorganisms. Compare time plots with normal curves. Microorganisms and fungus can compete with yeast for food. Microscope images and other tests required to isolate. Apply preventive measures to minimize bacteria or mold
Yeast inoculation	During fermentation: Smell Brix vs. time Temperature vs. time Specific gravity vs. time Post-fermentation: Sensory: Color off Sensory: Aromas off Sensory: Tastes off Microscope images	Metabolism and products of yeast vary with strain, temperature, oxygen, alcohol concentration, and available food. Refer to material section for yeast type considerations Compare brix, temperature vs. time curves with normal fermentation curve shapes
Nutrient deficiency	Pre-fermentation: test for YAN levels Track brix and temperature plots for healthy characteristics	If the yeast runs out of food and energy, it will shut down, stop metabolizing sugar to alcohol, and essentially become sediment. Different strains need varying levels of food (nitrogen) which guides level of food additions
Air	Post-fermentation: Sensory: Color off Sensory: Aromas off Sensory: Tastes off	Spoilage can occur when wine is exposed to oxygen. Acetic acid will grow rapidly with increasing temperature and oxygen
Dirty Equipment	Off wine characteristics may also indicate need to check cleanliness of equipment	Visual inspect equipment. Expect every use to potential leave residual contamination. Clean before and after use

In addition to your sense of smell, there are a many perceptive tests to aid in detecting VA that are shown in Table 5.7. These can and should be applied pre-, during, and post-fermentation.

Punch down or pump over is to prevent skins from drying out and growing microbes (which can cause high VA). The released CO_2 will hover above the caps because it is heavier than air—irrespective of punch down process.

Some yeasts are more tolerant to surviving within higher alcohol concentration musts. Most manufacturers will include alcohol tolerance of given commercial strains. Long experience, trial and error, and or testing is needed to assess natural yeast alcohol tolerances. Natural yeasts may provide more wine complexity but also are generally more unpredictable from year to year.

Even though sulfur dioxide is a material additive in many phases, it is an antimicrobial and antioxidant, and thus its effects are not a cause for VA. Refer to the environmental section for discussion of effects of temperature. Clean equipment and do not forget joints and all interconnecting tubing.

Table 5.8 is a summary of various preventive measures or actions you can take to reduce the likelihood of VA. Using good winemaking practice, frequent and timely tests, and consistent preventive measures is a big step toward improving the quality of your wine.

We have tabulated the likelihoods for this scenario's cause of VA in Table 5.9. At the end of chapter five, we will collect the assessments of the six major categories or bones, all of their sub-bones and rank order the mostly likely VA causes. Investigation of your test results and successful application of preventive measures should improve the wine. However, even the most disciplined proper fermentation

Table 5.8 Preventive measures that reduce fermentation probable causes to unlikely

Possible causes	Preventive measures	Prescriptive measures
Bacteria or mold	After crush, add standard SO_2 (by adding potassium metabisulfite) at 50 ppm	Measure free SO_2, pH, and determine addition to get to desired level for your condition. As part of process, also limit air exposure with wet caps via punch downs. Refer to environmental section for details.
Yeast. Affects vary with strain	Saccharomyces yeast is used for primary fermentation. Select type for varietal, alcohol tolerances, and provide for growth temperatures	Select yeast type and refer to material section for particulars
Nutrient deficiency	Understand nutrition needs and add appropriate nitrogen rich food	Select yeast type and refer to material section for particulars
Air	Apply punch down during fermentation which submerges dry cap within wine to keep wet and seal CO_2 in and air out	Punch down 3–4 times per day or continuously with pump over system
Dirty Equipment	Clean tanks and test equipment before and after each use	Adjust cleaning frequency before and after per procedure in upcoming Table 5.26

Table 5.9 Scenario 1 Red VA process phase 4 (fermentation) likelihoods

Possible causes	Likelihood for scenario 1	Comment
Bacteria or mold	Moderate	Sprayed vineyard in preceding step and added proper sulfur dioxide levels after fermentation, but not 100% effective
Yeast	Low	Used appropriate commercial yeast with proper inoculation methods
Nutrient deficiency	Low	Added prescribed food amounts at start of fermentation and checked multiple times
Air	Low	Punched down frequently and kept cap wet and air out
Dirty equipment	Low	Cleaned and sanitized all equipment contact surfaces

process will begin with a wide range of conditions and integrate many active ingredients. This confluence of marginal conditions pushes the fermentation process to be a leading suspect for causing volatile acidity. Many things need to go right at the same time. With perceptive tests, frequent sensing, and application of preventive measures, we are on our way to making great wine, but some unwanted microintruders may have slipped through.

We are going to jump to the aging process phase and investigate possible contributing factors to this scenario's VA problem.

5.3.4 Process Phase 6: Aging and VA

Aging in barrels is a relatively long process phase that will improve the character and style of your wine but can have unwanted outcomes due to unwanted products from fermentation, exposure to oxygen, or other instabilities in the wine. A good strong preventive measure is to test and add sulfur dioxide as needed.

VA may originate or be enhanced by inadvertent steps in aging. Table 5.10 lists possible causes, likelihoods, and comments about driving factors. Risks are increased due to the long time and potential risk of higher temperatures, exposure to oxygen, the presence of unwanted microbes, and contaminated barrels or tanks.

Aging wine may inadvertently expose the wine to oxygen. Fermentation was mostly oxygen free transitioning from grape juice to alcoholic wine with the large amount of CO_2 produced. Unwanted air may feed the small number of bacteria in the wine and spoil it. Topping off the barrels regularly and frequently prevents the build-up of air within the barrel. Not racking the wine and leaving it on the lees for long times can also allow residual unwanted bacteria to spoil the wine. The aging vessel may have unwanted bacteria on the surfaces or in the cracks between staves. Let us look at some tests identified in Table 5.11 that will aid in detecting some of these concerns.

Use your senses to look at color, smell the aromas, and taste the wine (Peynaud & Blouin 1983). These are good for finding faults. Our primary sensory test for

Table 5.10 High general likelihood of VA causes during aging diminished with preventive measures

Possible causes	Likelihood	Comment
Bacteria	High but low after PM	Bacteria in wine can cause instability and unwanted VA products
Air	High but low after PM	Growth rate enabled with oxygen. Air infuses headspace during evaporation
Too long on lees	Moderate but low after PM	Fermentation process should complete with dead yeast cells or other unwanted products remaining. Residual live yeast may continue fermenting unwanted VA products
Contaminated barrels or tanks	Moderate but low after PM	Barrel surfaces in contact with wine may be contaminated and introduce instabilities and unwanted products

Table 5.11 Tests to aid in detection of VA during aging

Possible causes	Tests	Comment
Bacteria	Sensory tests. Look for off colors, smell for off aromas, taste for off flavors frequently during aging	Bacteria produces acetic acid and volatile acidity
Air	Sensory tests. Look for off colors, smell for off aromas, taste for off flavors frequently during aging Inspection of headspace	This is approaching finished wine. Colors, aromas, and taste should be in line with your preferred style. Off colors, off aromas, and off flavors can all be produced by bacteria or mold. Growth varies with oxygen and temperatureNote size and rate of change of headspace. Oxygen will cause browning and enhance growth of unwanted products
Too long on lees	Audit status and history of wine and barrel use	Keep track of time on lees. New barrels provide most intense flavors and least chance of residual bacteria. Used barrels may contain imbedded bacteria
Contaminated barrels or tanks	Inspect, prepare, and clean barrels per contamination control plan	Clean before and after use

volatile acidity will be smelling off aromas. Well-balanced wines have many chemical elements including many volatile items and one group is volatile acidity. Check the aging barrels frequently and take note of the headspace as time passes and evaporation occurs. This will include some air and oxygen. Check your equipment for contamination, conduct lab tests of the wine to quantify the VA level. Compare trends or changes in VA level throughout the various winemaking steps. Audit your lab notes for history of barrels noting used barrels may containing bacteria and risk creation of VA products.

By applying preventive measures that are shown in Table 5.12, the risk of VA drops considerably. By applying preventive measures based upon the precise conditions of your wine, the risk drops even further.

Table 5.12 Application of preventive measures should always be applied during aging process

Possible causes	Preventive measures	Comments
Bacteria	Add sulfur dioxide to kill bacteria and not harm wine	Measure pH and free SO_2 and add sufficient SO_2
Air	Seal containers at start of aging. For porous barrels that allow evaporation, check headspace frequently. Top-off with inert gases and or wine.	Limit visual headspace
Too long on lees	Rack frequently during aging	Initial gross racking within 4 weeks start of aging. Rack every 3 months
Contaminated barrels or tanks	Clean barrel or tank inside contact surfaces before initial use and preceding racking operations	Clean and sanitize Refer to upcoming Table 5.26 for methods and procedures

For our scenario 1, Table 5.13 assigns the likelihood of each of top potential causes. During our aging process we racked but not enough and cleaned the barrels and equipment but not sufficiently. This increases the risk of air exposure and bacteria to the high and moderate levels.

We have made it through all of the process phases and have established various causes, likelihoods, and preventive measures for each. Table 5.14, the summary table of VA causes shows results for our scenario 1. We have dropped all of the low likelihood causes and kept the moderate and highly likely causes. It may truly require more specialized and expensive tests and analysis to isolate the one true root cause.

However, capturing the moderate and highly probable causes gives us a guide of how to apply our limited preventive measure resources to next year's wine. Due diligence and better adherence to the rules and discipline of known preventive measures could improve chances for better wine next year.

This section covered the first of six possible categories or "bones" of the fish. Congratulations, you have made it through the first bone or category, the process phases for our Scenario 1. Let us now move on to material for our red wine volatile acidity scenario one.

5.4   Material and Red Wine VA

We are now considering the next major bone on the road to root cause determination. This is the material bone. In general, wine flavors and aroma should come from the grapes without additives; however, winemaking does involve application of other materials which are limited by various Federal Regulations (Alcohol and Tobacco Tax and Trade Bureau, US Department of Treasury 2019). We are going to discuss the key material players including the grapes, yeast, yeast nutrients, and malolactic cultures. We discovered the excess VA just before bottling, so for this scenario, we are going to exclude the bottles or corks as suspects. Note that cork taint is common and should be investigated if you discover VA after bottling.

Table 5.13 Rank ordered likelihoods of excess VA caused during aging for scenario 1

Possible causes	Likelihood for scenario 1	Comment
Bacteria	High	Applied standard sulfur dioxide but did not test and calculate optimal amount
Air	High	Did not apply preventive measure of sufficient racking and topping off. Oxygen rich headspace too large and too long during aging
Too long on lees	Moderate	Late racking enhancing bacteria growth opportunities
Contaminated barrels or tanks	Low	Cleaned barrels before use and each racking

Table 5.14 Summary of most likely VA causes due to process for our scenario 1

Process phase	Possible causes	Scenario 1 high and moderate likelihoods
Aging	Bacteria	High
Aging	Air	High
Aging	Too long on lees	Moderate
Fermentation	Bacteria	Moderate

Fig. 5.9 Materials potentially impacting Red Wine VA

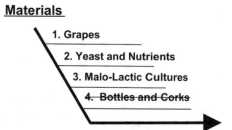

Volatile Acidity. Materials

Materials

1. Grapes

2. Yeast and Nutrients

3. Malo-Lactic Cultures

4. Bottles and Corks

5.4.1 Grape Materials

Figure 5.9 shows the material sub-bones of our winemaking fishbone. Volatile acids found in healthy wine grapes are generally considered negligible. The concentrations by weight are relatively small. However, the overall influence of this, albeit small level of acidity, is very important and critical. At harvest, the right levels of acids are critical, but their likely contribution to excess volatile acidity is very small. VA is generally a product of unwanted microbial metabolism and not the grape materials themselves. In the vineyard phase, as we discussed in the basic winemaking section, the key is to grow and harvest healthy grapes.

Note we have crossed off bottles and corks because for this scenario, the VA was discovered before that phase. Typical wine grape material composition by weight is shown in Fig. 5.10. Note how the acid percentage by weight is a relatively small number; however, our perception of acid in wine makes it very important.

Great wine with unique styles can be made from any healthy and sound wine grapes. The right level of acid, pH, and other components will provide for a truly balanced tasting wine. High acid concentration and low pH will also drive how sour

Fig. 5.10 Wine grape composition shows majority water and sugar and a very small percentage by weight of acid. However, acids are critical and our perception of acid in wine is complex and powerful

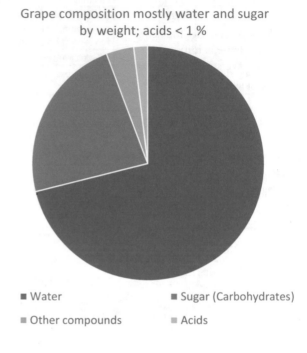

Grape composition mostly water and sugar by weight; acids < 1 %

■ Water   ■ Sugar (Carbohydrates)
■ Other compounds   ■ Acids

Fig. 5.11 Yeast goes into grape juice, and mostly CO_2 and ethanol comes out. Small concentrations of other compounds can have "oversized" influence on the quality of your wine

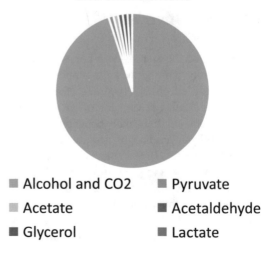

If you don't want out of balance levels after fermentation; take care of your yeast, food , and environment

■ Alcohol and CO2   ■ Pyruvate
■ Acetate   ■ Acetaldehyde
■ Glycerol   ■ Lactate

the wine tastes. We need the acid components in the grape; however, they are fixed and not a cause of excess volatile acidity. Figure 5.11 reminds us that there are many possible outputs from a fermentation if the conditions are out of balance and the yeast is not fed properly. The wine chemical composition and sensory tests will provide a baseline to track changes and trends. Refer to the basic winemaking and process bone section for preventive measures for growing healthy grapes.

For this scenario, the healthy grape component materials are not the cause of our excess volatile acidity as shown in Table 5.15. We use sulfur dioxide as a preventive measure. We still should employ tests listed in Table 5.16 to assess grape material health in case our conditions are extremely out of whack.

5.4.2   Yeast Material

We add yeast to start the fermentation. We chose to use a commercial yeast for better control; however, other experienced winemakers claim just as good control with a known vineyard and yeast that is "all around" their winery. The yeast is going to

Table 5.15 Grape component materials are fixed or low VA; however, bacteria can ride in on the skin

Possible causes	Likelihood	Comment
Water, sugar, and other components	none	Major components provide character but not inherent contributors to volatile acidity
Tartaric acid	none	Tartaric acid is a fixed, non-volatile acid that contributes to color and sour or tart taste but not aromas
Malic acid	none	Malic acid is also a fixed, non-volatile acid that contributes to color and sour or tart taste but not aromas
Citric acid	none	Citric acid is also a fixed, non-volatile acid that contributes to color and taste, but not aromas
Other volatile acids	Low	Concentration is generally too low to make any impact
Bacteria on skin	Moderate but low after PM	Use of sulfur dioxide inhibits bacterial growth

Table 5.16 Tests for healthy grapes, material quality, and upcoming decisions

Tests	Comment
Lab test grape must	Volatile acidity, soluble solids (brix), pH, titratable acidity (g/L), ammonia (mg N/L), free nitrogen (mg N/L), and yeast assimilable nitrogen (mg N/L) establish the pre-fermentation baseline
Sensory tests. Look, mold free, color, smell, and taste	Visual inspection, no mold. Acidity for intensity of sour taste. Recommend use of multiple judges to gather better statistics, averages, and throw-out outliers. This could be a part of planned test events at winemaking phases and including a broader range of aromas, mouthfeel, and overall quality

take grape ingredients and produce wine. Remember, the yeast itself comes with its own constraints on conditions. The winemaker needs to make sure the conditions are satisfied in order to get the desired products. One should not just add and forget. If you do not satisfy these other healthy yeast requirements, the "other end products" fraction in Fig. 5.12 will be large and consist of harmful stuff.

Saccharomyces is a genus or group of yeast types or species that are commonly used for beer, wine, and bread making. The output products are CO_2, ethyl alcohol, heat, and a few other items. The yeast needs food to work efficiently. The process is more effective at the right temperatures. This drives the need for temperature control of fermentation tanks or air-conditioned facilities. This is more critical with large volumes. There are various strains of Saccharomyces that have different characteristics.

Choose the yeast strains wisely. Yeast is available naturally from the vineyard and it is available commercially from many different suppliers. If you choose to go with natural yeast from the vineyard, it is moderately risky because of uncontrolled weather and adverse growing conditions. This may be less if there is a known predictable history of making good wine with yeast from the particular vineyard in questions. This wild yeast may also sneak into your winery by riding in on the skin

Fig. 5.12 Using particular yeast materials will subsequently dictate requirements on food and environment to prevent unwanted other end products

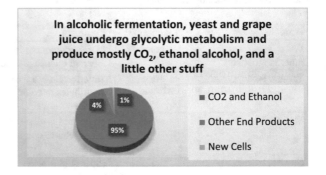

Table 5.17 Yeast can produce VA but with preventive measures, the likelihood of excess is low

Possible causes	Likelihood	Comment
Wild yeast characteristics	Moderate but low after apply Preventive Measure (PM)	Spoilage yeasts likely in vineyard, however, controlled with PM of addition of sulfur dioxide and limited exposure to air
Brettanomyces	Moderate but low after apply PM	Can be from vineyard on skin of grapes. Strong source of VA but controlled with addition of sulfur dioxide and limited air exposure
Commercial yeast characteristics	Moderate but low after apply PM	Majority of commercial yeast strains produce little volatile acidity with limited exposure to air
Winery yeast contamination	Moderate but low after apply PM	Spoilage yeasts may be present and thrive in a winery if cleaning processes are not practiced

of the grapes. If you want to go with commercial yeast, there are a number of options or choices to be made.

Table 5.17 is a list of potential fermentation and yeast causes for excess VA. They are similar likelihoods as long as you adhere to good winemaking practice and preventive measures.

Different strains allow the winemaker to influence certain winemaking conditions and styles. Choosing the proper yeast and controlling conditions can make for a good fermentation (UC Davis Extension Wine making certificate program. Linda Bisson, Grady Wann 2008). Table 5.18 is a useful checklist for choosing the type of yeast. Not paying attention to details, however, can cause slow or stuck fermentations.

Fermentation is required for winemaking but adds another risk for excess Volatile Acidity (VA). VA is a normal byproduct of fermentation with the very effective yeast, Saccharomyces. Excess VA is generally not a problem because it is created in small quantities. One notorious VA producer is the Brettanomyces yeast strain which can live on the skin of grapes, grow within wine, and can be a strong producer of acetic acid. However, it can be controlled with the addition of sulfur dioxide and a healthy fermentation cycle. Winemakers who avoid the use of sulfides are taking a high risk for microbial issues and high VA, so this would be a prominent root cause to flag.

For the industrious winemaker, there are yeast strains that produce relatively low VA. For this scenario one, the yeast type is very unlikely the root cause.

By virtue of using yeast we need to feed it. Add proper yeast nutrients. This is critical to productive yeast function and growth. A good test is to measure yeast assimilable nitrogen (YAN). Measure the must close to the desired start of fermentation. Higher than approximately 200 mg/L indicates a lower likelihood of a stuck

Table 5.18 Yeast selection checklist

Yeast selection checklist
1. Ability to ferment to completion with full dryness
2. Proper rate of fermentation: not too fast, not too slow
3. Wide temperature tolerance
4. Sulfur dioxide tolerance
5. Minimal off-character production
6. Maximize desired aroma characteristics
7. Function in presence of good microbes
8. Predictable fermentation characteristics
Follow manufacturer specifications for various yeast features to enhance fermentation and emphasize certain wine styles

Table 5.19 Tests that aid in diagnostics of material causes of VA

Tests	Comment
Volatile acidity level	Establish pre-ML baseline and compare post-ML levels
pH level	Make sure in range of 3.4–3.65 for best growth rate and malate metabolism
Sulfur dioxide concentration	Need approximately below 20 ppm for healthy ML fermentation
Nutrient addition amount and timing	May compete with yeast needs if before primary fermentation. Follow LAB specs and inoculant kits
Temperature	Make sure in range of 20 °C–37 °C. Growth better if higher end of range
Alcohol level	Measure level and should be <14%
Smell Inoculant	Off-odors indicate a bad starter culture
Acid composition	Measure for decrease in malic and increase in lactic for healthy ML
Wine aroma	Monitor for VA level changes
Wine taste	Taste for acidity reduction for healthy ML

fermentation. Increasing the YAN value can be accomplished with the addition of diammonium phosphate or "DAP."

Use tests (as shown in Table 5.19) to inform the next steps and follow the supplier's recipes to determine the appropriate amounts. Add close to the start of fermentation. Note that too much DAP and nitrogen can produce ethyl acetate and unwanted VA aromas. Too little nitrogen can cause other problems and a handful of other unwanted aromas such as H_2S or rotten egg smell. The takeaway is to stay within the acceptable range.

5.4.3 Malolactic Fermentation (MLF) with Lactic Acid Bacteria (LAB)

Malolactic fermentation reduces acidity, adds a little softer buttery taste, and can even improve wine stability. At the same time, some wine styles may not need less acidity nor more buttery taste. It is a winemaker choice. Figure 5.13 is a cartoon showing seven arrows pointing inward that can impact the process and three outward pointing arrows symbolizing possible results. Shades of green indicate mostly good, shades of yellow are a concern, and shades of red may be that path to unwanted VA.

ML can be easily implemented and managed but once the bacteria culture is introduced, it now becomes a requirement to manage it. Table 5.20 shows potential condition risks and general likelihoods of causing VA. Even with these new risks, excess VA can be prevented in a successful ML fermentation.

Testing provides objective data on the conditions of your wine. This is particularly helpful when adding new materials. This helps select the correct types, size the right amounts, and determine the best timing.

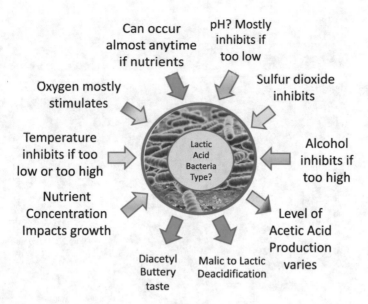

Fig. 5.13: Adding Lactic Acid Bacteria (LAB) material enables ML positives but brings new VA and other risks

Table 5.20 Adding bacteria introduces new possible causes but they are low after preventive measures

Add Oenococcus, new causes	Likelihood
Nutrient insufficiency	Moderate but low after add food
pH range too high	Moderate but low if test before inoculate
Temperature out of range	Low if measure and control
Incorrect sulfur dioxide addition	Low if test and add right amount of S_2O
Oxygen exposure	Moderate but low if limit exposure
Add Pediococcus or Lactobacillus	Likelihood
Above except problem if low pH	Moderate but low if test before inoculate

In the early days, we actually used a paper chromatography test to show the presence or absence of malic, lactic, tartaric, or citric in wine test samples versus control standards. An old photo of the results is shown in Fig. 5.14. Of course, this is not recommended for commercial wineries. Behind the garage it was time-consuming, yet fun, early method for determining if the ML fermentation was complete.

By following the preventive measures shown in Table 5.21, you should be able to reduce the likelihood of VA in your wine.

Our scenario one used Viniflora Oenos bacteria and the supplier recommended quantitative ranges. We did simultaneous primary and secondary fermentation on some varietals. On others, we completed primary fermentation first and then started secondary ML fermentation. Note that if you start ML after primary, one has to be vigilant about adding nutrients that might have all been consumed during the primary fermentation. Also note that some level of SO_2 is needed to combat unwanted

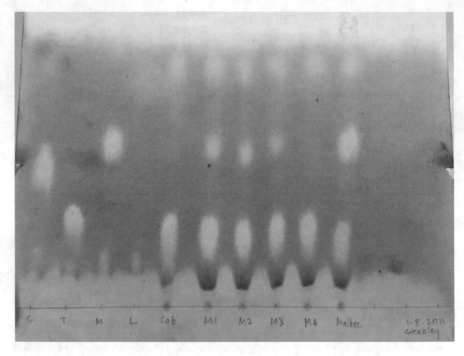

Fig. 5.14 Old school paper chromatography test to compare acid type levels to assess if ML fermentation is complete

Table 5.21 These preventive measures reduce likelihood of excess VA with LAB material addition

Possible causes	Preventive measures	Comments
Nutrient insufficiency	Add appropriate nutrients to feed bacteria	If concurrent with primary, sufficient for yeast, if after primary, sufficient for bacteria. Assess given volume and size accordingly
Correct pH range	Measure pH, choose appropriate LAB. If high adjust with tartaric addition. If low blend with higher pH wine	Suppliers specs for this scenario, ViniFlora Oenos ML Bacteria, show proper use for wine pH ≥ 3.2
Temperature out of range	Monitor temperature and keep warm enough to keep active. Warmth enhances activity	Suppliers specs for this scenario, ViniFlora Oenos ML Bacteria, show proper use for wine inoculation temperature 62–77 °F
Incorrect dosage of sulfur dioxide addition	Measure SO_2 and add as appropriate to kill unwanted microbes but not ML bacteria	Suppliers specs for this scenario, ViniFlora Oenos ML Bacteria, show proper use for wine inoculation max SO_2 of 40 ppm
Oxygen exposure	Keep containers sealed and top-off frequently	Check and top-off at least twice per month for typical 60-gallon barrels. Varies with barrel type and cooperage. In our scenario one example, we have added about 1/2 L every two weeks to a 225 L barrel

microbes, but a maximum must not be exceeded because that will kill the desired ML bacteria.

Table 5.22 summarizes the LAB addition risks for this scenario. The new risks are manageable and with the appropriate preventive measures, the additional potential causes are all low likelihoods.

We have completed this material bone for potential material causes of our excess volatile acidity. As shown in Table 5.23, we have determined that of each of the three major material groups, grape materials, yeast, and LAB are also all low likelihood. Our continuing mantra of applying preventive measures is once again critical.

5.5 Equipment and Red Wine VA

We do love our winemaking equipment. Figure 5.15 shows us the winemaking fishbone. Equipment makes wine production so much easier, but does it contribute to VA? Let us ask ourselves if the design is VA adverse. Does it make the risk of VA higher? From our process discussion, we can determine if some steps are worse than others. It is important to note contact extent, duration, and risk to spreading microbes.

If the equipment is misused, this is a generally a people training problem. Equipment material is generally inert and not a problem. If there is chemical reaction over long periods, like aging in an oak barrel, this is discussed in the material

Table 5.22 The risks from the material LAB are low for this scenario

Add Oenococcus, new causes	Likelihood
Nutrient insufficiency	Low after applying Preventative Maintenance (PM)
pH range too high	Low after applying PM
Temperature out of range	Low after applying PM
Incorrect sulfur dioxide addition	Low after applying PM
Oxygen exposure	Low after applying PM
Add Pediococcus or Lactobacillus	Likelihood
Above plus problem if low pH	Low after applying PM

Table 5.23 Summary: when applying preventive measures, our primary material additions are not likely to cause excess VA

Material addition summary	Likelihood	Comment
Grape materials	Low after applying PMs	Picked healthy grapes and followed preventive measures
Yeast	Low after applying PMs	Added yeast and nutrients, monitored conditions, sanitized winery, and followed PMs
LAB	Low after applying PMs	Added LAB and nutrient adjustments, and implemented PMs

Fig. 5.15 Equipment involved in making wine must come in contact with the grapes, juice, and wine

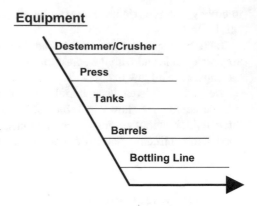

Table 5.24 Winemaking equipment has some inherent risk of enhancing excess VA

VA caused by equipment	Likelihood	Comment
Harvesting equipment	Low	The harvesting equipment itself is not a direct cause of VA if properly cleaned and you do not pick moldy damaged grapes
Destemmer/crusher	Low	Breaks skin and mixes skin and juice. Can transport unwanted microbes, but controlled with use of SO_2
Fermentation tank	Low	Fermentation tank is not a source of VA if properly cleaned. The duration of contact is also limited to fermentation period
Press	Low	Harsh squeeze of wine out of must. Low assuming cleaned and wine SO_2 concentration sufficient
Aging tank or barrel	Moderate	Moderate, even if cleaned, because evaporation in oak barrels and air leaks in joints of tanks. Long periods of exposure during long aging
Filtering and bottling equipment	Low	Direct filtering of larger particulate and microbes out of wine. Low assuming cleaned

section. However, barrels have a risky evaporation aspect we will discuss here. All possible causes can be controlled with preventive measures.

The destemmer/crusher and press have limited contact time with the juice; however, they have severe physical impact. The physical crushing of the grapes will break the skin and can mix unwanted microbes into the grape juice. This can be controlled with the addition of sulfur dioxide. The barrels are in very long contact with the wine but the barrel wood itself is at rest with the wine. Applying preventive measures such as cleaning and use of SO_2 can reduce the likelihoods of equipment related causes to low. It is important to maintain winery destemmer crusher, press, fermenters, and other winemaking equipment. These all contact the juice and wine and we need to assess if they are possible VA contributors.

Table 5.24 lists the equipment, general likelihood of causing VA, and background comments.

Visually inspect your equipment before purchase and during installation in your winery. It is amazing, so many smart phones these days have cameras. Figure 5.16

Fig. 5.16 Use oak barrels, top-off your wine!

Table 5.25 Visual inspection and functional operational checks before use can aid in assessments. Photographs can provide good records to assess changes

VA caused by equipment	Tests
Harvesting equipment	Inspect cutting edges for sharpness and cleanliness
Destemmer/crusher	Visual inspection, functional operation before use
Fermentation tank	Visual inspection for cleanliness, check logbooks for prior use
Press	Visual inspection, functional operation before use
Aging tank or barrel	Visual inspection for cleanliness, check logbooks for prior use
Filtering and Bottling equipment	Visual inspection, functional operation before use

is a picture of one of our barrels that we need to maintain a log of topping off and other maintenance tasks. We have written external codes on the barrel indicating purchase date and the wine inside. Take pictures of the equipment from all angles and good lighting. This establishes a baseline of pictures to compare changes, damage, and contamination.

Sometimes, a small camera phone can even allow good access behind and in between parts. Photograph critical parts that come in contact with the wine. Tests listed in the other sections may also trigger related VA investigation.

Table 5.26 Cleaning and sanitizing methods

Material	Equipment	Cleaning methods	Sanitizing methods
Glass	Bottles	Hot water rinse; sodium percarbonate soak if needed (chlorine-free oxidizing agent, 1 Tbls/gallon warm water); water rinse thoroughly	Rinse or spray with potassium metabisulfite solution (3 Tbls. potassium metabisulfite/gallon water), air dry
	Carboys	Hot water rinse; sodium percarbonate soak if needed; soft bristle brush; water rinse thoroughly	2 tsp. potassium metabisulfite and 1 tsp. citric acid/gallon water; cold water rinse, air dry
	Lab Glassware	Hot water rinse; sodium percarbonate soak if needed; water rinse thoroughly	Rinse with potassium metabisulfite solution; followed by distilled/deionized water rinse
Non-metallic	Bungs, O-rings, airlocks, etc.	Warm water soak, rinse or spray; rinse thoroughly	Rinse or spray with potassium metabisulfite solution
	Buckets, funnels, containers, etc.	Hot water rinse; sodium percarbonate soak if needed; water rinse thoroughly	Rinse or spray with potassium metabisulfite solution
	Fermenters/Tanks	High-pressure water power-wash	Hot water with sterilizing agent, and/or steam 12–20 min
	Hoses, racking wands, pumps	Warm water flow-through	Warm water rinse with sterilizing agent
Stainless	Destemmer/crusher	High-pressure water power-wash	Hot water with sterilizing agent, steam
	Filter	High-pressure water power-wash	Hot water with sterilizing agent, steam
	Clamps, fittings, racking wands, etc.	Hot water rinse; sodium percarbonate soak if needed; water rinse thoroughly	Rinse or spray with potassium metabisulfite solution
	Press	High-pressure water power-wash	Hot water with sterilizing agent, steam
	Tanks	High-pressure water power-wash	Rinse with 200-mg/L hydrochloride solution, citric acid rinse
	Bottling equipment	Warm water flow-through/rinse	Hot water with sterilizing agent, steam
Wood	Barrels	Hot water rinse, sodium percarbonate solution if needed	High-pressure water or high-power ultrasonics (Schmid, Grbin, Yap, & Jiranek 2011), steam 12 minutes (Cartwright, Glawe, & Edwards 2018)
	Press	Hot water rinse, sodium percarbonate solution if needed	Hot water with sterilizing agent, steam

Table 5.25 lists tests to aid in finding VA causing items. Equipment is generally not likely to cause or enhance excess VA if used properly and maintained. Oak barrels are one of the reigning champs for aiding in the production of great wine. For certain preferred styles, they can enhance desired flavors but with extended use they increase the risk of air exposure and excess VA. Cooperage of the barrel is important, but even the best bring some level of evaporation. This needs attentive frequent topping off. Timely addition of healthy wine to fill those air gaps controls oxidation and limits further growth of unwanted microbes. Even though it sounds repetitive, it is important, all of the equipment should be cleaned before and after each use. Table 5.26 is a comprehensive list of cleaning and sanitizing methods for various materials and equipment.

You probably have caught on already. Preventive measures are our "go to" strategy. It may sound simple, but it helps. Follow the manufacturers recommended maintenance plan. It is convenient to keep pre-mixed spray bottles of metabisulfite around the winery.

To prevent high VA occurring from winemaking equipment, follow the simple preventive measures listed in Table 5.27.

Assuming we have applied a disciplined preventive measures effort, we can prioritize our list of risky equipment causes of excess VA. The only one that makes it to the moderate level of risk shown in Table 5.28 is not topping off the oak barrels. The others can be considered low with cleaning and maintenance.

We have reviewed our equipment and now discuss our all-important people, the winemaking crew. Making great wine starts with great grapes from mother nature, but it is the care and attention of our team that brings out those wonderful aromas, flavors, and textures. Let us take a look at our people.

Table 5.27 Preventive measures are the key to our strategy

VA caused by equipment	Preventive measures
Harvesting equipment	Keep cutting edges sharp and clean before and after
Destemmer/crusher	Clean before and after use
Fermentation tank	Clean before and after use
Press	Clean before and after use
Aging tank or barrel	Clean before and after use. Sulfur toast inside of barrel before use. Top-off with healthy wine once or twice a month
Filtering and bottling equipment	Clean before and after use

Table 5.28 For our scenario one, we consider not topping off the barrels a moderate risk

VA caused by equipment	Likelihood	Comment
Aging tank or barrel	Moderate	Moderate, even if cleaned, because evaporation in oak barrels and air leaks in joints of tanks. Long periods of exposure during long aging

5.6   People and Red Wine VA

People can make great wine. Nature provides the ingredients, but people provide the recipes. Whether you are a commercial high-quality producer or a home winemaker (Fig. 5.17), people are involved. Operations will proceed well if they are trained and pay attention. They will also be more effective if your team is enthusiastic. Whether you want to produce a lot of quality wine or just a few high-quality cases, good plans, good instructions, good training, and good people make this possible. And of course, you are a good person for having bought and read this book!

Fig. 5.17 Mom's enthusiasm in the vineyard inspires the crew

Let us look at the actions of the people that may enhance unwanted excess VA. These people elements may originate with the vineyard crew, the winery staff, and/or the lab team. Figure 5.18 lists the people areas in our winemaking fishbone diagram. Note people are of course present in the vineyard, the winery production staff, and the winery laboratory staff. Table 5.29 indicates a number of people related causes and associated activities to help in identifying VA related issues.

Fig. 5.18 The people on your team will harvest the grapes, process them through the winery, and make critical lab measurements

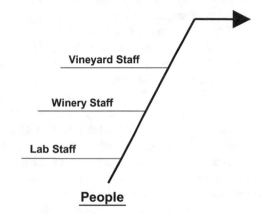

Vineyard Staff

Winery Staff

Lab Staff

People

Table 5.29 Likelihood of your team causing excess VA

Possible causes	Activities to aid in isolating people related VA causes	Comment
People related errors in vineyard	1. Audit plans, instructions, and training 2. Frequent joint vineyard talks and inspections and maintenance 3. Encourage crew to voice their observations and improvement ideas	Canopy management shortfalls or damaged grapes may overwhelm subsequent stabilization and microbe controls
Winery production team errors	1. Audit plans, instructions, and training 2. Observe screening operations bringing fruit into the winery 3. Production crew should maintain "juice to wine" logbooks and jointly audit results and trends	Logbooks may show missed sanitization, forgotten sulfur dioxide additions, or missed top-off. Attention to detail can be challenging with repetitive or routine tasks
Lab staff errors	1. Audit plans, instructions, and training. 2. Jointly audit logbooks of "juice to wine" treatments. 3. Jointly audit measurement result, note trends, flag inflection points, and highlight outliers	Unexplained out-of-family lab results may reveal measurement or result interpretation errors

Developing plans is a good beginning. For our scenario one, our vineyard, winery, and lab staff team have developed the plan and executed the plan (with help from family, neighbors, and friends). Even when many of the events are intended to execute critical winemaking operations, they can be fun and social. The effort needs to fit the situation. For example, the level of rigor and discipline for a commercial producer is very high but still needs to be rewarding for the people. Even for the home winemaker, enjoyment and fun are paramount, but you still must not allow errors to propagate into the quality of your wine.

Table 5.30 provides a list of training shortfalls that may contribute to excess VA. Plans and instructions may be in place but lack that critical detail step that could tip the overall process. Canopy management may understate the shady side of the vines or the harvest decision may come too late to get a trained crew on-site. Winery staff may not have the production experience to fill in the detail gaps in the instructions or procedures. The laboratory staff may forget to calibrate an instrument and add too little or too much. Check the quality of your plans and procedures and make sure the vineyard and winery staff are trained and up to date.

Train properly then turn your crew lose. Check in periodically. Routine, repetitive operations can often lead to distractions and lack of attention. Training, checking, and follow-up can keep the wine production on track, despite a chaotic world outside the walls of an efficient and effective winery.

When assessing this bone on the fishbone diagram we must determine whether the people are the root cause. So, if a detailed process exists but someone did not follow it properly, then it is a people problem.

We owe so much to the natural world and healthy grapes. We want to give them a rich environment to grow in. The same should apply to the care and feeding of our team! Figuratively and literally. Offer training in specialized skills and discipline in executing repetitive tasks. This is demanding and sometimes mundane, but always critical. It mostly works but make sure some of your winemaking is enjoyable. Table 5.31 is a preventive measures table but it is also a message to invest in your team to reduce VA and improve the quality of your wine.

Table 5.30 Activities to aid in isolating people training related shortfalls contributing to excess VA

Possible causes	Likelihood	Comment
Late harvest decision or poor training of picking crew	Moderate but low with careful preparation and training	Mistakes such as late harvest or picking damaged grapes may impact grape composition and quality
Winery production errors, inadequate equipment cleaning, or repetitive task errors	Moderate but low after training	Unwanted microbes may slip into winery and grow with missed sanitization, forgotten sulfur dioxide addition, or missed top-off
Lab staff errors in tests, sulfur additions, or sanitization oversights	Moderate but low with training and procedures	Inaccurate measurements may fail to provide early reveal of microbes, unstable wine, and increasing VA

Table 5.31   Invest in your team as you would in your vineyard and winery

VA caused by team	Preventive measures	Comment
People related errors in vineyard	1. Training in canopy management 2. Training in harvest 3. Keep logbooks before bud break and through harvest 4. Recognize team accomplishments	Educated, trained, and up to date crews make for healthy grapes and higher quality harvests
Winery production team errors	1. Training in production juice to wine end to end flow. Highlight microbe suppression from incoming grapes, fermentation and aging wine stabilization and test points 2. Keep logbooks and travelers with varietals and blends 3. Recognize accomplishments	Similar to above, well trained and diligent crew makes for better wine. Thorough logbooks allow for insightful investigations and continuous improvement
Lab staff errors	1. Training in wine chemistry 2. Focus on acid composition, sulfur dioxide concentration 3. Keep thorough accessible logbooks 4. Recognize accomplishments	Complex and accurate measurements benefit with additional specialized training. Maintaining logbooks allows for trending, discovery, and correlating errors

Fig. 5.19   Ask staff to keep logbooks or travelers for every vintage

Maintain thorough logbooks or travelers as shown in Fig. 5.19. These follow the grapes from the vineyard to juice in the winery to wine in the bottle. This keeps track of the wine progression. These records allow for review of past process steps and continuous improvement. They help correlate people errors that can be improved with additional training.

We leave the people section with one moderate group of possible likelihood of excess VA from winery production. Table 5.32 indicates that inadequate cleaning or repetitive task errors may be our problem. Better training is the most powerful preventive measures. Let us now take a look at possible causes in our measurements. Tasting and testing are of course very critical, but measurements must be right!

5.7   Measurement Errors and Red Wine VA

Let us consider measurement errors. Measurements are good but will still come with errors and uncertainty. They are not perfect or absolute. Could these errors result in excess VA in our wine? Let us look at Table 5.33 to see how we use the measurement information, how errors show up, and how errors impact treatments and the wine itself.

This discussion is about measurements and measurement error in your winemaking. Table 5.33 identifies various winemaking measurements and a commentary on the expected or typical associated errors. Please reference (Iland, Bruer, Edwards, Weeks, & Wilkes 2004) Patrick Iland's exceptional and practical discussion of wine measurement methods. It contains wine analysis concepts, test procedures, and trouble-shooting errors within the tests. For those who want a more mathematical and physical description of data reduction and error analysis, although not a wine focus, please reference Bevington (1969)[1].

Do not be mistaken; not making any measurements can lead to trouble. Figure 5.20 is a picture of a simple basket weighing measurement made at harvest. It is simple and quick. Many scales readily compute the tare weight by subtracting empty weight of the basket.

Measurements have errors but they can be kept small. With unknown problems in front of us, we still should review the various measurement error sources.

Table 5.32 Moderate likelihood of VA due to oxygen exposure due to winery staff error of infrequent top-off of wine during aging

Possible causes	Likelihood	Comment
Winery production errors, inadequate equipment cleaning, or repetitive task errors	Moderate with limited training. Our scenario detailed sustained attention. This created excess oxygen and microbial growth	Unwanted microbes may slip into winery and grow with missed sanitization, forgotten sulfur dioxide addition, or missed top-off

---

[1] This book might be considered far afield from enology and viticulture; however, it provides an excellent discussion of error analysis and is applicable.

Table 5.33 Measurement errors are present, but if typical, are unlikely cause of excess VA

Possible causes	Likelihood	Comment
Weight error	Low but not zero	Errors on the light side may occur in harvest grapes or measurement samples and/or additives and reduce stabilization effectivity
Volume error	Low but not zero	Errors on the light side may occur in harvest grapes or measurement samples and/or additives and reduce stabilization effectivity
Sugar concentration error	Low but not zero	High sugar musts might produce more VA. Erroneous low sugar estimates may misinform decisions
pH error	Low but not zero	Low pH wine, microbes, and oxygen may produce more acetic acid. pH measurement errors on the high side may misinform decisions
VA, TA, acid concentration error	Low but not zero	VA level is strong indicator of microbial action and instability. Low measurements may misinform decisions
Sulfur concentration error	Low but not zero	Key additive to stabilize wine but measurement error impacts accuracy of additive amount and changes with time

Fig. 5.20 Weigh grapes to guide the initial sulfite addition

Table 5.34 lists various aids to reduce or minimize the size of your measurement error and reduce uncertainty.

Figure 5.21 is our general winemaking fishbone subgroup associated with measurement errors. You may want to pause, have a nice glass of wine and a good night sleep, because we are about to take a mathematical measurement stroll through a typical wine stabilization activity. Even though measurement error is not a likely suspect for our excess VA, the following illustrates how errors come into play. Sulfur dioxide and its various forms are commonly used to preserve and stabilize wines. The molecular form of SO_2 is the most critical because it has antibacterial effects and helps prevent spoilage. The free form is the molecular SO_2 and this is the active form that kills microorganisms that can spoil wine.

Potassium metabisulfite is a relatively safe additive because it is stable and contains 57.6% available SO_2, the potassium does not impact the wine, and it can be easily measured by weight. Be sure to make pH measurements (Fig. 5.22) to

Table 5.34 Measurement errors are unlikely cause of our excess VA and making measurements and paying attention to detail always helps quality control

Measurement error	Preventive measures and aids to minimize measurement errors
Weight error	Calibrate with known masses. Repeat measurements throughout production. Reduce error with multiple data points
Volume error	Calibrate with known volumes. Repeat measurements throughout production. Reduce error with multiple data points
Sugar Concentration error	Collect random samples from wide variety of locations to give better representation of whole vineyard. Use independent measurement methods such as field refractometer, specific gravity test, different tests from different labs
pH error	Calibrate pH meter with known standards. Repeat measurement throughout wine production prior to additions
VA, TA, acid concentration error	Calibrate with known standards. Repeat measurement throughout wine production. Use different labs for different tests.
Sulfur concentration error	Calibrate with known standards. Repeat measurements to get more stats. Compare independent measurements from different labs

Fig. 5.21 Making great wine is aided by making measurements. Subsequent decisions may influence VA

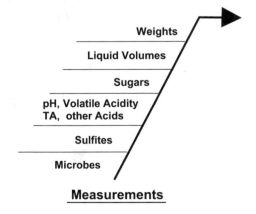

Weights

Liquid Volumes

Sugars

pH, Volatile Acidity
TA, other Acids

Sulfites

Microbes

Measurements

Fig. 5.22 This pH meter allows critical measurements with sufficient accuracy

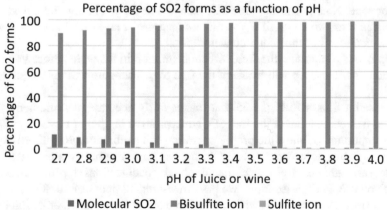

Fig. 5.23 The molecular SO_2 form is a very small fraction and still drops from 10% to <1% when pH increases from 2.7 to 3.8

understand your wine's pH and thus how it will react to additions and the molecular SO_2 form (Fig. 5.23).

It is also important to understand the health aspects of SO_2. Some individuals, such as chronic asthmatics are hypersensitive to SO_2. It is naturally produced by our bodies; however, excess amounts can have adverse reactions (but not headaches) in a limited number of particularly sensitive people. US wine labels are required to include the note "contains sulfites" to warn consumers. (TTB Alcohol and Tobacco Tax and Trade Bureau 2019)

We are now going to test our way through the pH, sulfur dioxide concentrations, and potassium metabisulfite ($K_2S_2O_5$) additions.

The following sequence uses a pH measurement of your wine, accounts for your production volume, and determines the appropriate addition of sulfur. We can use equation one to size the appropriate addition:

Equation one:

X ppm $\times$ 0.00657 $\times$ Y gallons = Z grams Where:

X = appropriate ppm of free SO_2 from trusted pH table
Y = volume in gallons of liquid
Z = grams of $K_2S_2O_5$ to add

The first step is to measure the pH of your wine. This is a measure of the concentration of free hydrogen ions. Some basics: the pH scale is from 0 to 14, plain water is 7.0, and most red wines are acidic and fall in the range of 3.3–3.6. We used a Hannah Instruments pH meter with a measurement accuracy of about 0.01 pH units. The Patrick Iland book on chemical analysis of wine recommends pH measurements with similar accuracies. (Iland et al. 2004)[2]

The pH is a big driver of the relative concentration of the various forms of SO_2 in juice or wine. Note the needed free SO_2 with increasing pH in Fig. 5.24. Also, note how big an error would result if you did not measure or know the pH of your wine, and planned to adjust molecular concentration.

Step two is to measure the free SO_2 concentration in the wine before you make any additions. This step will measure the free SO_2. There are various free SO_2 kits available.

We used a titrator in Fig. 5.25 that makes fairly accurate measurements. This provides a direct readout of the free SO_2 concentration in ppm or mg/L.

It is important to follow the proper operation and calibration procedures. For the low range it has an accuracy of 3% of the reading or approximately ±0.5 ppm, whichever is greater. Reference manufacturer specifications for the measuring device you use. For this example, we have measured 10 ppm of free SO_2.

In step three, we determine what level of free SO_2 we need to kill or inhibit bacterial or unwanted yeast growth. Refer to the exceptional book on winemaking theory and practice by Roger Boulton and the team at University of California, Davis.

---

[2] Modern equipment can provide improved accuracy and repeatability; however, it provides better insight to understand the basic methods as described in Chemical Analysis of grapes and wine: technique and concepts on P. 32

Fig. 5.24 Needed free SO₂ increases fast with increasing pH to achieve 0.825 mg/L molecular SO2 for protection

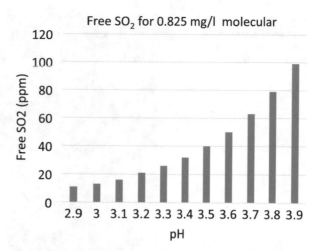

Fig. 5.25 This device titrates and provides an accurate SO₂ measurement

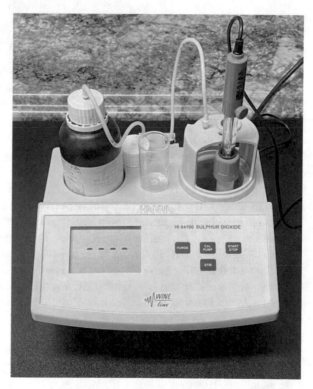

Fig. 5.26 Smaller scales with good accuracy and resolution can aid in precise additions

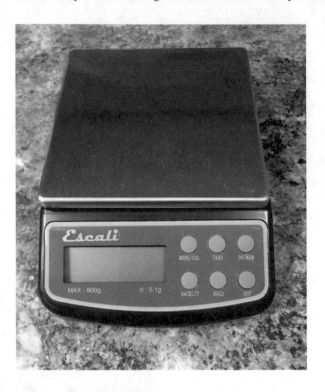

Boulton, Singleton, Bisson, & Kunkee (2010)[3] for comprehensive guidance. Higher levels of free SO_2 are needed as pH goes up. For this example, with a pH of 3.6, we need a free SO_2 level of 56.5 ppm.

In step four, we simply calculate how much to add by subtracting existing. Given that we measured 10 ppm in step 2, we only need an additional $56.5 - 10 = 46.5$ ppm.

In step five, let us figure out how much potassium metabisulfite we need to achieve the required free sulfur dioxide level. This scenario one example is for a nominal 60-gallon barrel. Referring back to equation one, our nominal calculation is 46.5 ppm × 0.00657 g/gal × 60 gal = 18.3 grams of potassium metabisulfite.

Errors may stack up and propagate their way through activities in different ways. Since we do not know exactly (almost by definition) what our errors might be, one can look at worst on worst conditions. If we were really having a bad day, all the errors could conspire to give us an unexpected or understated small addition. Or in the other extreme, they might all fall out to result in an unexpected large addition.

The errors shown in Table 5.35 reveal the surprisingly large magnitude of worst-case situations. However, it is not likely they will all stack up in a worst on worst-case manner. Most often, results will likely be somewhere in the middle. Note that these errors do not consider the full chemistry of the wine and the various strains of

---

[3] Note chapter 12 of referenced book for details on the role of sulfur dioxide and cell viability to be inhibited

Table 5.35 Worst on worst-case errors can add up

Step	Nominal truth	"Errored" smallest value	"Errored" largest value	Comment
Step 1: Measure pH	3.6	3.5	3.7	Our pH meter is ±0.01, but assume ±0.1 to be conservative representation of wide variety of meters and conditions
Step 2: Measure SO_2 (ppm)	10	9.7	10.3	Using 3% error on reading from our HI device
Step 3: Determine desired SO_2 (ppm) from studies	56.5	45.1	71.1	Based upon pH chemistry, uncertainty, and empirical results from studies
Step 4: Addition = Desired - existing (ppm)	46.5	34.8	61.4	Worst-case small is overestimate existing and underestimate desired. Worst-case large is underestimate existing and overestimate desired
Step 5a: Calculate potassium metabisulfite addition. Scale factor for unit conversion (grams/gallons)	0.00657	0.00657	0.00657	Assume insignificant error by using three significant digits
Step 5b: Measure volume of wine (gallons)	60	59.75	60.25	Assume ±0.25-gallon accuracy, volume measurements can be small
Step 5c: Final result: potassium metabisulfite addition (grams)	18.3	13.7	24.3	Note large worst-case extremes, if pH is off or not measured, could be worse

unwanted yeasts and bacteria, but this example does illustrate how each step may have its own error and they will convolve in some manner to impact the overall process. Consider the magnitude and accuracy of additions to find the right equipment or laboratory for the given measurement. Small scales (Fig. 5.26) can provide the right precision and/or accuracy you need.

For our scenario one, we have checked and found that we did maintain quite good laboratory discipline so even though we have measurement errors, we expect them to be small and also that we are not facing a worst on worst set of conditions. For this scenario one, we consider measurement errors are not a likely cause of our excess VA.

Congratulations, even though our measurement error did not likely cause our VA, we have covered a wide variety of measurement errors, ways to think about errors, and ways to keep them small. This knowledge will still help us make better wine. Let us press on with environmental effects.

Fig. 5.27 The
environment and potential
impacts on Red Wine VA

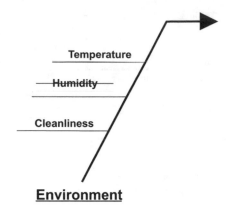

Temperature

~~Humidity~~

Cleanliness

Environment

Table 5.36 General environmental effects on causing excess should be investigated

General environment possible VA causes	Likelihood	Comment
Temperature	Moderate even with preventive measures	Most winemaking processes are impacted by temperature
Humidity	Moderate but reduced to low after preventive measures	Wet crushed fruit or dry corks might contribute but unlikely with good practice
Cleanliness	Moderate but reduced to low after preventive measures	Unclean and unsanitized facilities may contribute

5.8   Environment and Red Wine VA

We should investigate temperature, humidity, and cleanliness from our initial field
of environmental effects. Our general winemaking fishbone for the environment
group is shown in Fig. 5.27. We start by evaluating and crossing out humidity. This
is because our scenario's grape growing region is not excessively humid and does
not have unusually high levels of pests nor bacteria in the vineyard that thrive in
high humidity. In addition, low humid storage conditions and potentially dry corks
are also not applicable to this scenario. We also discovered the excess VA just before
bottling. General environmental effects that might cause excess VA or shown in
Table 5.36. We also assume typical preventive measures have been applied. If your
situation is different, adjust accordingly.

Cleanliness and contamination control are critical to prevent unwanted microorganisms entering the wine or winery (refer to cleaning and sanitization Table 5.26).
It may be a likely candidate for excess VA and must be checked in a problem investigation. For our scenario one, we have confirmed we likely had an adequate level
of cleanliness throughout the winemaking processes. This leaves us with
temperature.

A key part of our investigation is to follow the temperature conditions throughout
each winemaking process phase. Note the various grape growing and winemaking

phases in Table 5.37 and which have more or less temperature sensitivities. Table 5.38 provides a few recommendations on timing of temperature measurements. Fig. 5.28 is the type of digital thermometer we used; however, note there are many measurement technology and product options as your operation scales up and your need for volume, precision, and accuracy changes.

Vitis vinifera does not grow in regions with extreme cold winters and requires a sufficient number of hot summer days for the grapes to mature and fully ripen. Temperature will directly impact berry composition and growth. For this scenario we begin with healthy grapes grown in the proper conditions for the wine styles we desire. We may still be at risk for unexpected extreme weather conditions at the time of harvest. Rain at the time of harvest may promote some unwanted mildew growth. If we pick or damage the fruit entering the crush and we have hot temperatures, we may have combined conditions that together promote enhanced levels of unwanted bacteria.

Fig. 5.29 shows our vineyard weather station for monitoring temperature, rain, humidity, soil temperature, and soil moisture in our vineyard. Its control station is located in our winery and has a built-in temperature and relative humidity probe, allowing for continuous monitoring and recording of the winery environment too. We also use an inexpensive temperature monitor with a probe to test liquids during

Table 5.37 Temperature is a key driver in most winemaking operations and may contribute to excess VA

Focus on temperature as possible VA causes	Likelihood	Comment
Hot temperature along with damaged fruit	Moderate but reduced to low if kept cold	Microorganism activity increased in hotter temperatures. Enters winery on skin or damaged fruit
Hot temperature of crushed fruit	Moderate but reduced to low if kept cold	Hot temperature of crushed fruit increases skin extraction but also microbial growth
Hot extended maceration	Low if kept cold	If extended maceration gets too warm, this accelerates bacteria growth
Hot fermentation	Moderate, even with temperatures close to high end of tolerance	High temperature increases chemical rates. High temperature and oxygen increase microbial action.
Cold fermentation	Unlikely cause of excess VA	May cause stuck fermentation, but not likely cause excess VA
Temperature of press	Unlikely cause of excess VA	Not significant temperature dependent with mechanical presses
Hot aging temperature	Moderate but reduced to low if kept cold	Warm temperatures and oxygen for long period enhance microbial growth
Clarification or filtering temperature	Unlikely cause of excess VA	For our scenario, membrane pad filters, not significant temperature impact
Bottling temperature	Unlikely cause of excess VA	Not significant temperature dependence in short bottling operation
Bottle storage temperature	Moderate but reduced to low if kept cold	Warm temperatures and oxygen for long periods enhance microbial growth

Table 5.38 Temperature monitoring and frequency guidance

Focus on temperature as possible VA causes	Temperature monitoring
Hot temperature along with damaged fruit	Local weather stations on most phone apps. One °C accuracy sufficient
Hot temperature of crushed fruit	Liquid digital thermometer
Pre-fermentation maceration[a]	Liquid digital thermometer frequently. Daily
Hot fermentation (Ough, Amerine, & Sparkls 1969)	Liquid digital thermometer frequently. 3x Daily
Cold fermentation	Liquid digital thermometer frequently. 3x Daily
Temperature of Press	Outdoor/indoor thermometer
Hot aging temperature (Boulton et al. 2010)	Facility continuous temperature monitor
Clarification or filtering temperature	Facility temperature monitor
Bottling temperature	Facility temperature monitor
Bottle storage temperature	Storage room continuous temperature monitor

[a]Boulton et al. (2010)

Fig. 5.28 Digital thermometers are convenient for measuring winery temperature during fermentation

fermentation and punch down. These are just a few of the many examples of methods and tools to measure temperature during various phases.

Table 5.39 lists various temperature control aids that reduce the likelihood of VA and improve the quality of your wine. Find the approach that best fits your situation.

After our investigation, we have confirmed the right temperature control precautions were in place. Table 5.40 shows for this scenario, the likelihood of temperature causing our high VA to be unlikely. It was easier for us because of the small volumes; however, these methods are scalable to larger examples.

Fig. 5.29 Vineyard weather station is good for temperature monitoring throughout growing season. But without extremes, vineyard temperatures are not likely cause of excess VA

5.9 RCA Summary of Red Wine VA

We have searched for the cause of our excess volatile acidity. We have investigated the world of possible candidates. This world was categorized by process, materials, equipment, people, measurements, and environment. We determined which items were likely and which were not. We prioritized respective possibilities. Table 5.41 is a summary of the likelihood of various winemaking phases that cause our VA problem.

Table 5.39 Take precautions to control temperature throughout entire winemaking process

Possible temperature excess VA risks	Temperature control aids
Hot temperature along with damaged fruit	Pick fruit early in the morning for cooler temps. Discard damaged fruit
Hot temperature of crushed fruit	Inspect and discard damaged fruit before crushing. Crush in the shade and or air-conditioned facility.
Cold soak	Before fermentation, keep temperature below 55°F to limit yeast action. A/C room
Hot fermentation	During fermentation, the punch down keeps cap wet, but also releases heat.
Temperature of press	Perform in shade or A/C room
Hot aging temperature	Age in temperature-controlled room. If you are lucky and have the resources, caves are cold and good!
Bottle storage temperature	Age in temperature-controlled room.

Table 5.40 This scenario's excess volatility is unlikely to have been caused by too hot temperatures

Focus on temperature as possible VA causes	Likelihood	Comment
Hot fermentation	Unlikely	This scenario's fermentation temperatures were kept cold
Hot aging temperature	Unlikely	This scenario's fermentation temperatures were kept cold

Table 5.41 Prioritized causes by moderate and high likelihood help guide preventive measures

VA caused by process	Likelihood	Comment
Aging	High	Aging with bacteria and air present highly likely cause of excess VA
Aging	Moderate	Too long on lees from lack of racking contributed to bacteria growth
Fermentation	Moderate	unwanted bacteria grew during fermentation
VA caused by equipment	Likelihood	Comment
Aging tank or barrel	Moderate	Moderate, even if cleaned, because evaporation in oak barrels and air leaks in joints of tanks. Long periods of exposure during long aging
VA caused by people	Likelihood	Comment
Winery staff production errors	Moderate	Given presence of bacteria and oxygen in aging barrel, failure to top-off contributed to excess VA

Congratulations! We have made it through this excess VA chapter and we been able to identify excess VA moderate likelihood causes for our scenario. No, we have not isolated and identified a single root cause; however, we have identified contributing factors and preventive measures. Even though the one true root cause may be difficult to isolate, preventive measures can be applied to next year's vintage! Yes, it will take dedication, informed work, and time. It is not easy and there is no simple recipe, it very complex. Now you have a powerful set of tools. You are well on your way to "Quality wine, every time."

References

Alcohol and Tobacco Tax and Trade Bureau, US Department of Treasury. (2019). Alcohol regulations. Retrieved from TTB: Alcohol and Tobacco Tax and Trade Bureau: https://www.ttb.gov/other/regulations.shtml#alcohol

Bevington, P. R. (1969). *Data reduction and error analysis for the physical sciences*. New York: McGraw-Hill Book Company.

Boulton, R. B., Singleton, V. L., Bisson, L. F., & Kunkee, R. E. (2010). *Principles and practices of winemaking*. Davis, CA: Springer Science +Business Media Inc..

Cartwright, Z. M., Glawe, D. A., & Edwards, C. G. (2018). Reduction of Brettanomyces bruxellensis populations from Oak barrel staves using steam. *American Journal of Enology and Viticulture, 69*, 400–409.

Iland, P., Bruer, N., Edwards, G., Weeks, S., & Wilkes, E. (2004). *Chemical analysis of grapes and wine: techniques and concepts*. Adelaide: Patrick Iland Wine Production PTY LTD.

Ough, C. S., Amerine, M. A., & Sparkls, T. C. (1969). Studies with controlled fermentations. Xl. FERMENTATION TEMPERATURE EFFECTS ON ACIDITY AND pH.

Peynaud, E., & Blouin, J. (1983). The taste of wine: The art and science of wine appreciation. (M. Schuster, Trans.) Paris: John Wiley and Sons, Inc.

Schmid, F., Grbin, P., Yap, A., & Jiranek, V. (2011). Relative efficacy of high-pressure hot water and high-pressure ultrasonics for wine oak barrel sanitization. *American Journal of Enology and Viticulture*, 519–526.

TTB Alcohol and Tobacco Tax and Trade Bureau. (2019). Sulfite mandatory labeling items. Retrieved from TTB.gov: https://www.ttb.gov/wine-resource-tool/section07-labeling/sub-section1/71120.htm

UC Davis Extension Wine making certificate program, Bisson, L., & Wann, G. (2008). *Wine production for distant learners. lesson 8: yeast biology, choice of yeast strain and nutrition management*. Davis, CA: UC Davis Extension.

University of California Agriculture and Natural Resources. (2019). How to manage pests: Grape.. Retrieved from UC Integrated Pest Management Program: http://ipm.ucanr.edu/PMG/selectnewpest.grapes.html

Chapter 6
Red Wine Imbalance Problem Solving with Cause and Effect Analysis

This chapter will address red wine imbalance. A wine may taste too boozy with too much alcohol and not balanced with other components. When serving most food items, one can also serve salt and pepper on the table or when serving coffee and tea, one can serve cream and sugar. This allows your guests to add either to balance and fit their own personal tastes. With wine, the balance is set by the grapes and wine-making process. You can and should pair the right wine with the right food; however, when serving the wine, it is too late to adjust its balance.

We start with defining what we mean by imbalance, then discuss our imbalance problem statement, and apply the cause and effect diagram. We will identify possible causes and assign likelihoods. We will point out key tests to help guide us into a better understanding of what is really going on and discuss preventive measures. Finally, we will organize and prioritize the possible causes. As we mentioned in the prior scenarios, this process will first widen the number of possibilities and hopefully catch them all. Then it will narrow the field and prioritizing a list of contributing factors. We will work toward identifying our true root cause. Figure 6.1 shows a nice-looking vineyard. The message is that balance begins in the vineyard; however, you as a winemaker, may lose it, keep it about the same, or maybe even improve it.

We proceed to sort these candidate drivers into priority order to focus our time and resources on resolving the leading candidates.

Alcohol and acid are both wonderful components in a great tasting wine; however, too much alcohol can taste hot and boozy or too little acid can leave a wine flabby and watery. There is a proper balance. Pulp maturity may be when the pulp sugar-acid balance is just right. However, do not forget the skins and seeds. The skins and seeds contribute to color, astringency, and other aroma and flavor balance in a properly made wine. Our challenge is to find the imbalance causes amongst all of these elements that are working for or against one another. Maybe we need to enhance one item and limit another …at the same time. This RCA method will guide us through a process to find the problem and lead us closer to a quality balanced wine. Again, one of our keys is to not waste precious resources on unlikely

© The Editor(s) (if applicable) and The Author(s), under exclusive license to
Springer Nature Switzerland AG 2020
J. Steakley, B. Steakley, *A Quest for Quality Wine, Every Time*,
https://doi.org/10.1007/978-3-030-34000-1_6

Fig. 6.1 Balance can begin in the vineyard. Throughout vinification; you might lose it, or you might keep it, or you can even improve it!

causes. With good practice and application of preventive measures, we will make better and better wine.

6.1   Red Wine Imbalance Basics

We are striving for quality wine. It is a joy to anticipate, a joy to experience, and a joy to remember. The word quality itself has so many different meanings to so many people and so many different applications (reference Chap. 2). But for winemaking, for this book, and for this chapter, we are going to focus on one particular aspect of quality wine, and that is balance. As part of root cause analysis, our scenario for this chapter is a "wine without balance." It sounds as bad as it tastes.

We are going to think about balance in terms of various components which should be in harmony with one another. Which components? Figure 6.2 shows the relative fraction of pulp, skins, and seeds in a grape berry. There are subcomponents within these components. How much of each drives balance? Better balance is a key part of a higher quality wine. No single element stands out nor clashes with its neighbors. Many wine faults or marginal winemaking practices can impact balance. Imbalance lowers quality and ruins what might have been a wonderful opportunity.

We are driving hard toward balanced wine by designing the vineyard and wine-making process to consider multiple components together. These components need to be in the correct relative concentration with respect to one another. The challenge will be figuring out a reliable measure of that. We will draw upon grape biology, wine chemistry, and sensory assessments. This gives you a shot at better chances to improve the wine, vintage after vintage. Table 6.1 lists the key elements we will track in search of better balance.

We will consider balancing varietals vs. climate, yield vs. ripeness, and sugar vs. acid. After harvest, we move through the vinification process phases evaluating sugar, alcohol, acid, fruit, and tannin. By considering all of these together and inter-preting their relative levels, we improve our chances of making balanced wine. Table 6.2 comments on these elements in vinification.

Judge and balance various components against each other, against standards, and changing trends. Make informed and balanced vinification decisions.

Fig. 6.2 Understanding what makes up the grape components early gives us a better chance of making a balanced wine at the end

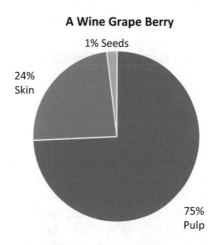

Table 6.1 To achieve quality and balance, we will take an integrated view of key elements from the vineyard and throughout the winemaking process

Elements	Comment
Sugar	Sugar develops as the grapes ripen. Most sugar is converted to alcohol during fermentation. Residual sugar may remain at various levels depending upon the wine style
Alcohol	Alcohol is produced by yeast during fermentation. Alcohol can produce a hot burning sensation when high and also can have a sweet flavor
Acid	Acid decreases as the grapes ripen. Choosing the harvest date allows for optimizing the balance between sugar, acid, and fruit
Fruit	Wine grape varietals each have their own unique fruity aromas and flavors. Different complex styles can be made from single varietals or blends
Tannin	Tannins are a phenol and come from the skins and seeds. Tannins play a role in aging, body, and mouthfeel
Water	Even though water is the largest component of wine, wines character and complexity are driven by the absolute and relative concentration of the above elements

Table 6.2 Consider multiple components and their relative concentrations and balance in winemaking decisions

Vinification focus elements	Comments
Sugar	Sugar is converted to alcohol during fermentation and residual sugar is near zero for dry red wine
Alcohol	Alcohol is produced by yeast during fermentation. Alcohol can produce a hot burning sensation and also can have a sweet flavor
Acid	Acid will produce a crisp, refreshing taste. Too much may produce out-of-balance heavy sour taste
Fruit	Wine grape varietals each have their own unique fruity aromas and flavors. Do not mask fruit flavors with out-of-balance vinification impacts
Tannin	Tannins are a phenol and come from the skins and seeds. Tannins play a role in aging, body, and mouthfeel

6.2   Imbalanced Wine Problem Statement and Cause and Effect Diagram

We have discussed our concept of balance and are ready to write down the problem statement and effects for this chapter's scenario (Fig. 6.3). Capturing the facts of your "imbalanced" situation and objectives allow you to move through the winemaking phases and fishbone covering key potential causes. Our scenario in this chapter seeks to fix imbalance in our dry red wine.

Our problem does not indicate a major fault but leaves us wanting better balance. The color is a nice red but there is little fruit in the aroma. The taste is pleasant but the hint of blackberry is slight and offset by a heavy vegetative flavor. The sugar at harvest was high but the acid was somewhat low.

The fishbone (Fig. 6.4) is tailored for our imbalance situation and the timing of our observations. We still have some aspect of each of the six bones, process,

Fig. 6.3 Our imbalance
investigation begins with a
thorough problem
statement

- Dry Red Wine
- Color is red garnet
- Aroma has no glaring faults
- Taste of a little red berry
- Not quite full-bodied in taste
- Don't taste any residual sugar
- Noticed imbalance after aging and before clarification ad bottling
- 2015 Cabernet Sauvignon grapes
- Grown in Santa Cruz Mts. AVA
- Sugar 27°Brix at harvest
- Titratable acidity 6.8 g/L at harvest
- pH 3.4
- Volatile acidity 0.1 g/L at harvest

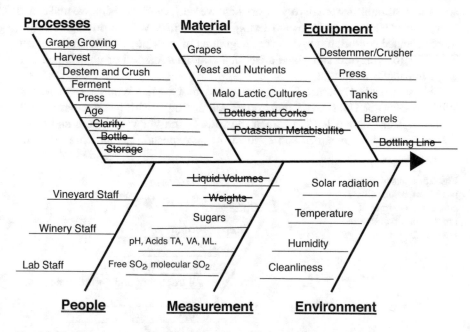

Fig. 6.4 Fishbone is tailored for imbalanced wine problem statement

material, equipment, people, measurement, or environment that may have caused imbalance in our wine. Let us investigate each.

6.3   Process Phases and Imbalanced Red Wine Problem

We will now dive into the process phases of winemaking shown in Fig. 6.5.

We have crossed out the process phases that occur after the timing of our observation of imbalance. Note that in other potential imbalance situations, clarification may be a contributor to imbalance. It is also possible but very unlikely that storage might be a cause for imbalance. However, for this chapter's scenario, we discovered discernible imbalance before clarification.

6.3.1   Grape Growing and Harvest Process Phases

We are faced with an imbalanced situation. The concept of this root cause analysis scenario is straightforward. Every assessment we start, every decision we make, and every action we take work toward fixing an imbalance. We are driving toward a more balanced condition from the very beginning in the vineyard. Carefully review Table 6.3 for vineyard and grape areas that may have caused or contributed to a grape imbalance at harvest.

This scenario's Cabernet Sauvignon grows best in the Winkler (Amerine & Winkler, 1944) or UC Davis Summation scale[1] region II with just enough warmth or degree hours above 10 °C to ripen the fruit. Is the vineyard layout and trellis design arranged for maximal sun exposure? Sun exposure has an extraordinarily

Fig. 6.5 Isolated process phases can be initially culled out by virtue of problem observations and likelihoods

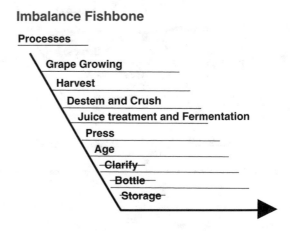

---

[1]The Winkler Index or scale classifies grape growing regions by heat summation or growing degree days.

Table 6.3 Likelihood of an imbalanced wine can begin with an imbalanced vineyard

Grape and vineyard element	Imbalance likelihood	Comment
Grape varietals appropriate to growing region	Moderate	If at installation or replant stage, match varietal type to local soil and climate conditions
Vineyard layout and trellis design	Moderate	Orientation, spacing, and trellising can enable good sun exposure, efficient canopy management, improve growth, and provide better quality berries
Overwatering	Moderate	Heavily watering can increase yield, but over-watered and plump grapes reduce concentration and intensity of flavors. Deficit irrigation can maintain health and improve quality
Inadequate pruning and/or canopy management	Moderate	Pruning and canopy management requires frequent attention. Possible to achieve high yields from your vineyard without overcropping. Provide best environments for vine growth and berry microclimates
Sugar/acid balance at harvest	Low	Sugar increases as the grapes ripen, but acid decreases. Late harvest imbalanced grapes destined for dry red wines can result in high alcohol but low acid, flabby wines

strong effect on the balance of sugars, acids, and other soluble solids within the grape berries well before harvest (Crippen & Morrison, 1986). Was the canopy management optimal for fruit growth? Heavy watering may have increased yield but compromised flavor intensity. Were the grapes picked too late giving a high sugar to acid imbalanced ratio? Table 6.3 lists and comments on vineyard elements and associated likelihoods of causing imbalance.

The tests listed in Table 6.4 can aid in providing objective evidence to help the grape grower and winemaker understand the vineyard situation.

Review the climate history and test the vineyard soils. If you are fortunate to have a grape growing history at the vineyard or nearby neighbor's vineyards, understand their prior growth experiences. Inspect the vineyard orientation and trellis design for optimal arrangement and sun exposure.

Check with local weather stations and if available, humidity and soil moisture sensors to confirm adequacy of watering. Figure 6.6 shows a shaded hillside vineyard that has the challenge of reduced sun in the mornings. Review canopy management practice and frequency. Audit the ratio of sugar to acid and trends approaching harvest and prior vintages.

Table 6.5 lists preventive measures or actions you take in the vineyard that will help maximize balanced growth and result in ripe flavorful grapes. Matching the varietal to the growing region is good if you are starting out. Even if you already are working with a planted vineyard, knowing the climate data and tested soil conditions will guide the extent of vineyard management.

Deficit irrigation can help find the minimum yet sufficient watering levels to grow healthy and flavorful grapes and still achieve plentiful yields. These techniques have been used and studied (Mathews & Anderson, 1988) and proven to be beneficial. Water deficit methods can be implemented before and or after veraison

Table 6.4 Tests results provide objective evidence for areas to correct imbalance in the vineyard

Grape and vineyard element	Tests	Comment
Grape varietals appropriate to growing region	Research historical local climates, nearby weather station data. Test soil composition. Consult varietal characteristics	Note the Winkler index based upon the concept that grapes grow best with a given number of hours a day of temperatures above 10 °C. Vines and grapes metabolize more efficiently at warmer temperatures. Soil conditions provide for water and nutrient availability
Vineyard layout and trellis design	Sun angle orientation and trellis shape, row spacing, and in-line vine spacing. Maximize hours of sun on vines	Orientation, spacing, and trellising can enable good sun exposure, efficient canopy management, improve growth, and provide better quality berries
Overwatering	Research local rainfall. Soil moisture sensors or nearby weather station data	Heavy watering can increase yield, but over-watered and plump grapes reduce concentration and intensity of flavors. Deficit irrigation can maintain health and improve quality
Inadequate pruning and/or canopy management	Sugar, color, and taste of grapes on vines. Size and number of clusters per vine. Yield-to-pruning ratio	Pruning and canopy management requires frequent attention. Possible to achieve high yields from your vineyard without overcropping. Provide best environments for vine growth and berry microclimates
Sugar/acid balance at harvest	Brix and acid concentrations at harvest. Fruit to vegetative grape flavors	Sugar increases as the grapes ripen, but acid decreases. Late harvest imbalanced grapes destined for dry red wines can result in high alcohol but low acid, flabby wines

to achieve different effects and maintain or improve balance in the vineyard. Table 6.6 indicates some data from two vintages in our own vineyard. Notice how much the balance or ratio of sugar to acid changed from 2016 to 2017. You should compare against quality standards but also remember to understand the data assumptions it represents, such as the regional extent where the vineyard sits, growing region, AVA, state or country.

Trellis design and canopy management are very important for balance in the vineyard. There is a delightful quote from Galileo Galilei: "Wine is sunlight held together by water." Hearing this always brings a smile. But the working winemaker and vineyard manager need to dig into wine grape biology, soils, irrigation, and canopy management. There is an exceptional resource for canopy management, "Sunlight into Wine" (Smart & Robinson, 1991). A second thorough reference for managing your vineyard is Soil, Irrigation, and Nutrition (Nicolas, 2004). The vineyard manager can bring theory and affordable practice to maintain balance in the vineyard.

Fig. 6.6 Hillside vineyards maybe challenged by shade and low morning sun. Thin neighboring trees, check spacing, and confirm growing season temperatures adequate for varietals

Consider Berry Sensory analysis (Winter, Whiting, & Rousseau, 2004)[2] to provide direct sensory assessment of your berry's skin, pulp, and seed characteristics to better inform on ripeness and readiness to pick.

It is always exciting, if not nerve-racking, how much things change year to year and week to week. This is a good argument for making more checks, testing more parameters, and comparing more ratios. Keep a record of grape characteristics and their relative balance to compare with finished wine sensory characteristics and its relative balance.

We have the vineyard and harvest possibilities to deal with. Table 6.7 shows us the likelihood of imbalance originating in the vineyard and harvest. We believe our vineyard site climate and soil are not optimal for growing Cabernet Sauvignon. However, there are many viticulture methods and techniques to make the most of this site. We will strive, year after year, to get the most fruit and least veggie balance we can. Let us move on to the next process phase, destem and crush.

---

[2] When you are in the vineyard and far from the lab, you can learn so much from the berry's look, taste, and feel. Use your own senses more of the time.

Table 6.5 Preventive measures accomplish balanced grapes begin with grape varietal selection, need attention throughout the growing season, and end with harvest timing

Grape and vineyard element	Preventive measures	Comment
Grape varietals appropriate to growing region	Match varietals with climates. Cabernet sauvignon is late ripening, needs warmer, drier environment. Hearty vine that grows in most soils with good drainage	Note the Winkler index based upon the concept that grapes grow best with so many hours a day of temperatures above 10 °C. Vines and grapes metabolize more efficiently at warmer temperatures. Soil conditions provide for water and nutrient availability
Vineyard layout and trellis design	Vertical shoot profile trellis designs provide for good sun on vertical wall of leaves and grapes. Good height and access for canopy management	Orientation, spacing, and trellising can enable good sun exposure, efficient canopy management, improve growth, and provide better quality berries
Overwatering	Deficit irrigation is a must. Still enable healthy grape maturation, but limit excess vine growth	Heavy watering can increase yield, but over-watered and plump grapes reduce concentrations and intensity of flavors. Deficit irrigation can maintain health and improve quality
Inadequate pruning and/or canopy management	Provide winter pruning and spring, and summer thinning for adequate but minimal shoots and leaves. Provide pre-veraison good cluster sun exposure. Inspect vineyard weekly	Pruning and canopy management requires frequent attention. Possible to achieve high yields from your vineyard without overcropping. Provide best environments for vine growth and berry microclimates
Sugar/acid/fruit balance at harvest	Brix and acid concentrations at harvest. Fruit to vegetative grape flavors	Sugar increases as the grapes ripen, but acid decreases. Late harvested imbalanced grapes destined for dry red wines can result in high alcohol but low acid flabby wines

6.3.2   Balance Impact from Destem and Crush and Cold Maceration Juice Treatment

The destem and crush process phase breaks the grape skins and allows the skin, pulp, and seeds to mix together. This operation is a necessary part of winemaking and will integrate and blend the qualities of the skin and grape berry juice together. We do not consider this process a source of imbalance as described in Table 6.8, but an activity to blend grape skin, pulp, and seeds together. Any resulting imbalance after this process phase is a result of the qualities of the incoming materials. There is an option to skip destemming and crushing and go right into whole cluster fermentation. This might be appropriate for some varietals and styles; however, it is not recommended for Cabernet Sauvignon because of its inherent relatively high concentration of pyrazine. This produces a noticeable vegetal taste. We are trying to lower and balance the vegetal to fruity taste.

Table 6.6 Monitor your own vineyard and climate situation carefully

Year	Cabernet sauvignon	Pre-harvest	Harvest day
2016	Date	9/21/2016	10/3/2016
	Sugar	22.5	23.1
	Acid	7.7	6.42
	pH	3.16	3.47
	Sugar/acid ratio	2.9	3.6
	S/A ratio change in 12 days	Baseline	23.1%
2017	Date	10/4/2017	10/8/2017
	Sugar	22.0	23
	Acid	8.3	5.85
	pH	3.4	3.64
	Sugar/acid ratio	2.7	3.9
	S/A ratio change in 4 days	Baseline	48.3%

Data from other sources is important but understand the assumptions behind it. Big changes in values and balance can occur faster than you might think

Table 6.7 We do have a range of contributing factors and possible causes for an imbalanced wine originating in the vineyard

Grape and vineyard element	Likelihood of imbalance cause	Comment
Grape varietals appropriate to growing region	Moderate	In the real world, there are not always good affordable options on where to plant. For our scenario, we have a cool climate and clay loam soil. This will challenge getting fruity and well ripening Cabernet Sauvignon
Vineyard layout and trellis design	Low to moderate	Given our site, our vineyard has employed good VSP methods and orientation with respect to the hillside and sun
Overwatering	Low to moderate	Our CA site has been in a drought situation for a number of years; however, we have delayed irrigation but probably could take even stronger water deficit measures
Inadequate pruning and or canopy management	Moderate to high	Our vineyard canopy management has been good, but more thinning and leaf removal will reduce vegetative flavors in the wine
Sugar/acid/fruit balance at harvest	Low to moderate	Threatening rains and a volunteer harvest crew may push harvest days away from the optimum

Before fermentation, a winemaker may choose extended cold maceration to get more color and taste. Cold maceration allows more time for the skin and juice to mix and react. We are going to ferment and press pretty soon and these are fairly quick operations, so there is good merit in just letting the skins and juice blend together for a few more days before you start fermentation. If your grape skins have the right characteristics, this is a very effective way to infuse color and flavor into the juice. Keeping it cold limits start of fermentation. The next process phase to impact balance is fermentation.

Table 6.8 Destem and crush are not a likely source of imbalance for our cabernet sauvignon scenario 2

Destem and crush and cold maceration	Balance impact likelihood	Comment
Crush and destem clusters and berries	Low	Most destem and crush methods do a relatively efficient job of crushing the grape and allowing a mix of juice
Skip it all together. No destem and crush. Go directly to fermentation	Moderate to high but NA if not used	We do not recommend skipping the crush destem operation for Cabernet Sauvignon which already has a high risk of vegetal tastes due to high pyrazine concentrations
Cold maceration post crush and pre-fermentation	Moderate to high but NA if not used	This additional time for skin and juice contact time will definitely change the balance of the wine. This is a phase that can really improve wine color and flavors

Fig. 6.7 Even though you start with balanced grapes, an improper fermentation may introduce imbalance

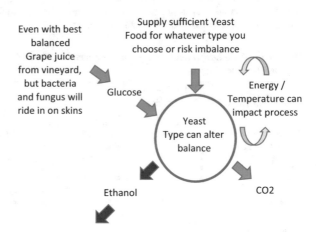

6.3.3   Finding Wine Balance, or Imbalance Problems in the Fermentation Process Phase

This scenario is about finding the cause of our imbalanced wine. Figure 6.7 illustrates inputs, outputs, and key drivers of fermentation. You have considered the grape growing, harvest, destem and crush phases, now let us consider potential imbalance effects from fermentation. Imbalance can be caused by elements in all six categories: process, materials, equipment, people, measurements, and environment. We might have had balanced grapes from the vineyard; however, there is always the possibility of bacteria and or fungus riding in on the skins. Such organisms may contribute to imbalance in the wine from this process.

Fermentation is that very critical winemaking process that converts our grape juice and other miscellaneous "hitchhikers" to alcoholic wine. Yeast is our principal catalyst. At the same time, it brings imbalance risk as shown in Table 6.9. There is considerable activity in this microscopic world that will likely change the balance of the juice and wine alcoholic wine.

Table 6.10 lists various tests that can aid in achieving balance or finding imbalance problems. We will defer yeast type and temperature discussion until the upcoming materials and environment sections. Bacteria or fungus in the presence of the air will interfere with the fermentation process. Improper inoculation of the yeast may also introduce an imbalance cause. Not feeding your yeast enough nor often enough will inhibit yeast growth.

Let us conduct various tests and inspections to help us discern higher risks and likelihoods for imbalance in our process. Even if not all these tests can be conducted, each one provides an additional improvement in your understanding of the actual condition of your wine. Consider employing services from local winemaking laboratories or testing services to obtain more comprehensive wine chemistry analysis. Figure 6.8 shows us an example report for our 2016 vintage.

Comparing and analyzing before, during, and after test results will provide clues to balance and imbalance conditions. We are in this for the long term, we expect to continue year after year, and should compare vintage to vintage results.

We are not allowed to chaptalize (add sugar) in making wine in the USA; however, we can add acid if the levels are too low. By measuring and tracking the acid levels from the grapes, to juice, and the post-fermentation wine process phase, you have the data in order to determine if the acid is too low and if the balance needs adjustment.

Fermentation presents many possible causes of imbalance but planning with preventive measures as shown in Table 6.11 and monitoring will play a big part in balanced winemaking. If something unexpected occurs, such as a very rapid fermentation cycle with little time on the skins and the color is light, but the aromas and flavors are still good, one might even consider a style change for this particular

Table 6.9 Fermentation transforms juice to wine, but unfortunately it also adds an imbalance risk

Possible causes	Likelihood	Comment
Unwanted bacteria or mold	Moderate but low after PM	Bacteria and molds can cause unknown effects and impact the balance, for good or bad. If you like the style, malolactic fermentation can be from a good bacteria and good process that will lower the acid balance of the wine
Yeast inoculation	Moderate but low after PM	A healthy fermentation should start with a proper inoculation. Winemakers should prepare the culture per a trusted supplier's specifications. Will cover the temperatures later in the environment section
Nutrient deficiency	Moderate but low after PM	Malnourished yeast can result in a stuck or unhealthy process. This will produce unwanted byproducts that will negatively impact the sensory balance of the wine
Air	Moderate but low after PM	The growth rate of bacteria, a potential source of imbalance, can increase with oxygen. Keep the cap wet and the air kept out

Table 6.10 Tests made before, during, and after fermentation can provide information on potential cause of imbalance

Possible causes	Tests	Comment
Bacteria or mold	Before and after wine composition tests Before and after sensory characteristic tests Temperature and specific gravity vs time during process	A good baseline with wine composition and sensory characteristics allows identification of changes that matter. This provides a better picture of what matters to you the winemaker and how to make a difference
Yeast inoculation	Check inoculate for large quantities of yeast cells. Smell for off aromas of inoculant	Large number of yeast cells (several million per milliliter) drive rapid domination over microorganisms. Prepare inoculant per supplier's recommendation. Blend with wine sample before adding preparation to full batch
Nutrient deficiency	Pre-fermentation: Test for YAN levels Brix and temperature plots for healthy process progress	Stuck or slowed features may indicate insufficient nutrients if the yeast runs out of food and energy, it will shut down, stop metabolizing sugar to alcohol, and essentially become sediment and clearly impact balance of wine
Air	Visually inspect cap for wetness and coverage. Punch-down cap frequently throughout fermentation	Spoilage can occur when wine is exposed to oxygen. Acetic acid and imbalance will grow rapidly with increasing temperature and oxygen

Fig. 6.8 Your lab or local winemaking shops may provide affordable component test capabilities

Sample Information	9/21/2016
Client:	Steakley, Joyce
Vintage:	2016
Variety:	Cab Sauv
Vineyard:	Steakley
Must Panel Results	
Wine pH:	3.16
Titratable Acidity (TA) g/l:	7.69
Titratable Acidity (TA) in %:	0.77
Sugar in Brix:	22.5
Alpha Amino Acid in mgh/l:	86
Ammonia mg/l:	126
Yeast Available Nitrogen mg/l"	185
Volatile Acidity g/l:	0.1
Giuconic Acid g/l:	0.0
Malic Acid g/l	2.6

Table 6.11 Imbalance likelihood from fermentation can be reduced with the right preventive measures

Possible causes	Preventive measures	Comment
Bacteria or mold	Add potassium metabisulfite to kill bacteria. Limit air to prevent bacteria growth	Refer to Chap. 5.7 measurement section for SO$_2$ addition formulas. Do not add if ML secondary fermentation is active
Yeast inoculation	Inoculate with large numbers of yeast cells. Prepare inoculation solution per supplier's instructions. Prefer solution base is sample of grape juice ready for fermentation	Large numbers of yeast cells help rapid domination of desired yeast over other unwanted organisms. Inoculant solution should otherwise be as similar as possible to target grape juice and temperature. Refer to material section for yeast strain selection
Nutrient deficiency	Understand nutrient needs of selected yeast and with pre-fermentation YAN tests and add foods accordingly. Maintain awareness of healthy fermentation by tracking sugar concentrations and temperature	Refer to material section for additional comments on yeast foods for particular yeast types
Air	Apply punch down during fermentation to submerge dry cap in wine to keep cap wet and air/oxygen out	During fermentation, less air is better. Punch down 3 to 4 times per day or daily with pump over system

Table 6.12 Bacteria or mold and exposure to air remain moderate likelihood potential causes of our imbalanced wine for this scenario 2

Possible causes	Likelihood to cause imbalance	Comment for this scenario
Bacteria or mold	Moderate	We took reasonable precautions to clean and sanitize winery and equipment; however, unwanted bacteria or fungus could have breached our fermentation process
Yeast inoculation	Low	We followed recommended rehydration of the yeast with correct water volume and temperature
Nutrient deficiency	Low	YAN levels in must were good. We added recommended food supplements per yeast specifications and monitored fermentation progress through to completion
Air	Moderate	We punched down 2–3 time per day; however, there may have been a few days where we only punched down once.

batch. Some yeasts are more aggressive and fast acting than others. We will discuss this more in the material section. Experimenting with the production line may not be practical for a large operation trying to achieve a long-established historical style and customer preferences, however, do not be afraid to pull aside small batches for a little R&D and experimentation in alternative winemaking methods.

The presence of bacteria or mold along with exposure to air is the combination that can ruin a healthy fermentation. Table 6.12 identifies these as our moderate likelihood causes for imbalance in this scenario. This effect has an unpredictable outcome of fermentation products and consequently and unpredictable impact on imbalance. We must work to prevent and avoid this all together.

In general, we are striving for a healthy fermentation process, as opposed to steering it toward influencing or changing the balance. By virtue of conducting a healthy fermentation, we will have the best chance of achieving a balanced wine. Note however, we will discuss wine balance characteristics influenced by yeast type in the materials section coming up later. The next process phase in our basic wine-making sequence is the press phase.

6.3.4 Imbalance Impact from the Press Phase

The pressing process phase squeezes the juice out of the must or what remains of the crushed grapes and fermented wine. This is usually done with equipment and does not add material. It does, however, move more skin components into the wine. We have used a hydraulic press which consists of an interior balloon that fills with water and expands. This expansion pushes or squeezes the crushed grapes and fermented wines up against a metal screen. The wine flows out and the pressed material is discarded.

There is a possibility of pressing a second time to get a little more wine out. Under a second press or even a very heavy initial press force, one might squeeze more unwanted components from the skins and change the balance of the wine. Most often this introduces bitter compounds and is not recommended. The most natural wine comes from the "free run" which is the wine that flows out from the fermented must without any press action or force, other than gravity. Proper use of the press will make this an unlikely source of imbalance. We do not believe the press process phase is a strong candidate for the cause of the imbalance in our wine, so we will transition immediately on to the next process phase of aging.

6.3.5 Finding Wine Imbalance During the Aging Process Phase

Aging is the long-sustained process of maturing the fermented wine. If you are using barrels, aging too long will cause imbalance with barrel flavors overpowering the wine fruit. However, aging too long may not actually occur that often in the commercial market, because there is strong financial motivation to get the wine finished and to market sooner. Aging too quickly shortens time to infuse barrel components into the wine. This may unbalance certain wine styles that prefer barrel aged oak complexities. Table 6.13 comments on possible causes of imbalance due to aging and their associated likelihoods.

Table 6.13 Possible causes of imbalance due to aging

Possible causes	Likelihood	Comment
Aging too long or short	Low to moderate but low after preventive measures	Too long causes imbalance of barrel flavors dominating over fruit. Too short fails to infuse oak and may unbalance certain wine styles
New vs. old barrels	Low to moderate but low after preventive measures	New oak barrels infuse wine faster. Used oak barrels may have little effect. Best duration shortens if new and lengthens if used
Air exposure	Low to moderate but low after preventive measures	Air infuses headspace during evaporation. Growth rate of bacteria enabled with oxygen. Air causes imbalance of color by browning and cause bitter flavors
Bacteria, too long on the lees, or contaminated barrels or tanks	High but low after PM	These elements introduce unknown and unpredictable changes to your wine. They will most likely cause imbalance

New oak barrels are fresh and impart heavier oak components in shorter times. After use, the barrels lose their components and become "neutral" aging tanks. The higher level of oak element concentration is reduced through migration into prior wines. The barrels themselves will not introduce major changes, but they may not improve the complexity that you, the winemaker, might have desired. Certain finished wine styles are achieved with long aging in oak barrels. Used oak barrels may not, or at least will take longer, to introduce that vanilla smoky flavor that some nice balanced wines are known for.

The items in Table 6.13 within the row labeled "Bacteria, too long on the lees, or contaminated barrels or tanks", are all grouped together because they all can create unknown conditions. Unknowns make your wine characteristics unpredictable. This means one cannot reliably predict how the finished wine will look, smell, or taste. This will most likely result in unbalanced wine. The best option is to avoid this all together by paying attention to test results and use best practice and preventive measures.

Fortunately, we have tests and various actions that can aid in heading off this unpredictable path toward imbalanced wine due to aging. These various test types and actions are listed in Table 6.14. Making chemical composition tests and sensory tests frequently are a powerful tool. We are going to coordinate with the periodic need to top-off the evaporating wine from within the oak barrel to establish a periodic rhythm to aid in a periodic sampling scheme. Whatever equipment and process you use; complementary processes help the synergy and reliability of getting them done.

It is very difficult to establish absolute and universal levels that are good or bad thresholds or go, no go decision values. However, it is possible and practical to establish relative balance levels for your own particular wine and personal tastes. This does take time and commitment but it will improve your chances for a balanced wine, every time. Noting your own trends for a particular vintage and comparing vintage to vintage provides you with your own standards.

Table 6.14 Tests and aids to diagnose imbalance due to the aging process

Possible causes	Tests	Comment
Aging too long or short	When topping off wine, make chemical composition and sensory evaluations. Note sensory impressions of fruitiness vs oak flavors and aromas, and complexities	Track changing conditions. Build trend for each vintage and note timing of balanced state for your particular wine and style preferences. It takes diligence
New vs old barrels	Decide on preferred style, your budget, and available timeline	New barrels impact strong oaky compounds faster and longer. Old used barrels are more neutral; however, use saves money. Prioritize your driving goals
Air exposure	Each barrel or container type is unique and may allow wine evaporation and increasing air gap. Inspect periodically to understand air gap size and rate of change. Set periodic calendar reminders to top-off wine barrels	Inspect for increasing air gaps within barrels, not just size at the moment to guide top-off amount, but rate of change to help predict next top-off date
Bacteria, too long on the lees, or contaminated barrels or tanks	Inspect, prepare, and clean barrels per contamination control plan	Clean and sanitize before and after each use. Also, clean and sanitize winery surroundings

There is a general trend for relative fruitiness to fall and woody complexities to rise while aging in oak barrels. We are trying to follow this changing balance closely. We also want to understand how long we should age our wines. Sample the barrel frequently to provide periodic sensory and or chemical tests. Check the color against a white uniform background to note any changes. If you commercially sell your wine, then there are clearly cost considerations for aging and holding the wine longer in your winery. For this imbalance example, we are going to focus only on the quality and sensory style of your wine. The goal is to achieve a balanced wine in accordance with your own preferred style.

Polyphenols in wine are responsible for its color. Wood tannins are phenolic compounds. They dissolve fairly quickly in wine. Wood barrels allow for the constant penetration of oxygen through the staves and into the wine. The tannins in the aging wine are themselves oxidized. These tannins influence or temper the process. The oxygen might otherwise be directly oxidized. The tannin reaction produces a more crimson color (Vivas & Glories, 1996). The presence of these tannins also prevents the growth of the brick-yellow color. Your wine conditions, of course, may be different. This is why you taste and test all throughout the process.

We can learn and understand so much about our wine by how it smells. We can use our sense of smell to get to know the aromas and how our wine is maturing. We created Table 6.15 based upon the work of Ann Noble and the Wine Aroma Wheel (Noble et al., 1987).

Table 6.15 Identify unique aromas and how they change

Primary group	Subgroup	Unique aroma
Fruity	Citrus	Grapefruit
		Lemon
	Berry	Blackberry
		Raspberry
		Strawberry
		Black currant (cassis)
	Tree fruit	Cherry
		Apricot
		Peach
		Apple
	Tropical fruit	Pineapple
		Melon
		Banana
	Dried fruit	Strawberry jam
		Raisin
		Prune
		Fig
	Other	Artificial fruit
		Methyl anthranilate
Woody	Burned	Smokey
		Burnt toast
		Coffee
	Phenolic	Medicinal
		Phenolic
		Bacon
	Resinous	Oak
		Cedar
		Vanilla

The aroma wheel enables communicating and thinking about aromas in a way that is more precise and understood by others. With practice, it also makes it easier for you to remember and track what is important to your own style of wine. Each of the grape varietals brings characteristic aromas and flavors. The vineyard terroir is also imparting aroma and flavor characteristics. We have fermented, pressed, and are now aging. We are going to use this type of breakdown to track the evolving characteristics of our wine.

This chapter focuses on using the periodic sensory evaluations of color, aroma, and taste to guide us toward better balance. This information also allows us to ferret out flaws that may emerge during aging. Two major relevant groups are fruity and woody. Generally, on the first day of aging, there is probably a lot of "fruity" and not much "woody." On the last day of aging it probably has reversed in that there may be relatively less "fruity" and much "woodier."

The balance will have changed. We are going to use this changing trend to help us decide when enough aging is enough. It may also be obvious, but this is a timely reminder, "fruit" is coming from the grape origins of the wine and "woody" is coming from the oak barrel. We would like to break these down into even more detailed and familiar descriptors so we can more accurately track sensory attributes that rise and fall.

If the categories are too big and unfamiliar, they may get lost in the shuffle. As time progresses, changing wine characteristics can introduce imbalance. After sampling the barrel during aging, smell the wine and get to know the aromas. By following unique wine features, you the winemaker, get more information to decide if you like the wine as it is. That is time to stop barrel aging and bottle it! Practical considerations may adjust that moment by weeks or a few months until the bottling line is ready.

This, of course, is subjective and varies with each individual. It depends on personal perception capabilities and style preferences. Tracking multiple features helps give better insight into different emerging trends. Practice also helps improve your own capabilities. This is a "Two for One," it improves both your wine and winemaking skills at the same time!

Cabernet Sauvignon has a fruity aroma[3] with berry characteristics that range from blackberry, raspberry, strawberry, and black currant. Since we are barrel aging in this scenario, we are also looking for wood imparted wood characteristics. These are broken down into the unique aroma column of smoky, burnt toast, coffee, medicinal, phenolic, bacon, oak, cedar, and vanilla. Note that some of these, like vanilla, or smoky are commonly considered good in quality wines. However, some of these like medicinal are not good.

By noting your own wine characteristics throughout the wine aging process, you can understand the evolving sensory balance. This can indicate when your wine is vectoring off to an unwanted imbalance. Tracking perceived changes in aroma and flavor characteristics will help when imbalance occurs and also provide useful guidance on when to stop aging and move onto clarification and bottling.

It is helpful to keep a record of your own sensory perceptions. This can be qualitative or semi-quantitative. It helps if it is done in a way that allows you to determine if it is going up or going down. Is it getting more intense or less intense? This could be a three-level scale like low, medium, and high or a numerical scale of one to ten. Whatever way works best for you. We prefer a numerical scale of 1 to 10 where one is the lowest and 10 is the highest. It may take some time to get calibrated to your own senses; however, there are simple spreadsheets to track the trends.

Table 6.16 is an example of fruity and woody aromas over 18 months. The rows are unique aromas and the columns are various months after the wine goes into the

---

[3] The table subgroup and unique aromas are aroma characteristics that were developed by Ann Noble for the Aroma Wheel.

barrel. The absolute magnitude of the numbers is not as important as how the intensity of the aroma notes and thus the scoring goes up or down with time in the barrel. This table shows a typical trend of stable or slightly decreasing relative fruity notes vs increasing woody aromas in our Cabernet Sauvignon barrel aged wine. The trend also provides an indicator as to when the wine is achieving the balance and style of your preferred wine. This trend can help guide when you should plan to stop aging and prepare for clarification and bottling. These sensory tests can also indicate when and to what extent imbalance starts to occur.

Similar evaluation and scoring can be performed to follow changes in flavor trends. Figure 6.9 shows an example trend. Multiple data points are gathered throughout the aging process based upon periodic tasting. My goodness, it is not like we need another reason to taste more wine! These tests and aids provide information to guide preventive measures. Let us discuss what these are for the aging process phase.

During aging, we have identified possible causes of imbalance, the length of time, using new or old barrels, exposure to air, bacteria, and possible length of time on the lees. The preventive measure table identifies various actions to reduce these risks.

Review and use the sensory and chemistry aging test results to guide length of time. This is one of the key preventive measures to establish proper balance. These timely tests also answer the length of time question for either old or new barrels. It is the balance in the wine that you are evaluating. If you are also limiting the air exposure, this aging process should proceed properly. Proper topping of the aging container will prevent this. Do not allow any or very little air space inside the barrel.

Table 6.16 Keep a record of perceived intensity values of fruity and woody aromas in aging wine

Unique aromas				
Fruity aromas	*Month 1*	*Month 6*	*Month 12*	*Month 18*
Blackberry	6	6	6	6
Raspberry	3	2	2	0
Strawberry	2	2	2	2
Black currant (cassis)	7	6	6	6
Average fruity	4.5	4	4	3.5
Woody aromas	*Month 1*	*Month 6*	*Month 12*	*Month 18*
Smokey	0	2	2	2
Burnt toast	0	0	2	2
Coffee	0	0	2	2
Medicinal	0	0	0	2
Phenolic	0	0	0	0
Bacon	0	0	0	0
Oak	0	2	4	6
Cedar	0	0	0	0
Vanilla	0	2	4	6
Average woody	0	0.667	1.556	2.222

Table 6.17 Preventive measures during aging are part of best practice to reduce risk of imbalanced wine

Possible causes	Preventive measures	Comments
Aging too long or short	Using your frequent sensory and or chemical tests, look for trends that fit your style preference	If you want or need more woody flavors, age longer in your barrels or add oak chips or staves. Stop aging earlier for more relatively intense fruity flavors
New vs. old barrels	Plan in accordance with style preference. If budget precludes new barrels, age longer for higher woody characteristics	Use frequent tests to guide length of aging
Air exposure	Seal containers at start of aging. For porous barrels that allow evaporation, check headspace frequently. Top-off with wine if possible. Use inert gasses as necessary	Check frequently enough to limit evaporation and headspace to typically less than 0.1–0.2 gallon
Bacteria, too long on the lees, or contaminated barrels or tanks	Clean barrel or tank insides before and after use. Rack frequently. Clean container every racking before refill with wine	Initial gross racking within 4 weeks start of aging. Rack at least every 4 months. Refer to Table 5.26 for winery cleaning and sanitization procedures

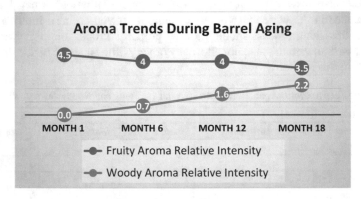

Fig. 6.9 Sensory evaluations during aging can indicate trends

Proper cleaning of the containers, before and after use, is also a great way to limit the unwanted bacteria population and protect the wine.

Table 6.17 provides a list of preventive measures and actions that can reduce the risk of imbalance being introduced or worsened during aging.

Given these aging imbalance risks, tests, and preventive measures, let us now review what the most likely causes of imbalance are for this scenario in Table 6.18. We will work very hard at the true best practice; however, we will not be perfect.

Aging is a wonderful process phase with the wine in direct contact with our wood barrel for a long time. This unfortunately still leaves us with a number of possible imbalance risks for this scenario shown in Table 6.19. It is good that we also have a number of tools and preventive measures at our disposal. In this scenario, we discovered the imbalance at the end of the aging process and before clarification. That means for this section, that is the end of the investigation for the process phase bone. Now, let us summarize the moderate and high possibilities for imbalance due to process.

The high, moderate, low-moderate, and low possibilities for the process phase of our imbalanced wine are tabulated. The grape growing conditions are highly likely; however, there are other possibilities. Even with knowing recommended preventive measures, we may have imperfect application or effectivity of preventive measures. One or some combination of these may have allowed imbalance conditions to emerge. Vigilance and discipline will continue to improve next year's vintage. The principal material source in our wine is the grape itself. Let us take a look at the grape materials and other incidental materials introduced during the winemaking process.

6.4   Materials and Possible Causes of Red Wine Imbalance

In the basic winemaking discussion in Chap. 5, we noted the importance of picking the right grape varietal for the growing region, choose the harvest date for ripe grapes, and basic healthy growing practices. We also discussed methods to test, isolate, and prevent potential causes in the viticulture and grape growing phase. In Chap. 5, we noted the grape composition and relevance to our volatile acidity scenario. In this scenario we have sensed an imbalance that seems like too much veggie and too little fruit. Figure 6.10 is our materials bone for this imbalanced wine scenario. We smelled and tasted the imbalance after aging but before corking and bottling. Within the United States AVAs, we are not allowed to add flavor components in the winemaking process so the mature grape components are the focus. Wine is often referred to by the simple expression, "Wine is fermented grape juice." Grape juice comes from vineyard grapes and yeast enables fermentation. So, let us illustrate our RCA method by investigating the materials in our grapes and yeast.

Table 6.18 Process summary of possible imbalance

Possible causes	Likelihood	Comments
Aging too long or short	Moderate	We did perform sensory and chemical tests, but we could have more frequently and regularly
Air exposure	Moderate	We did check the barrels and topped off a number of times, however, here again, we sometimes used an inert gas and we did not top-off with wine often enough
Bacteria, too long on the lees, or contaminated barrels or tanks	Moderate	We did clean, but did not properly sanitize the used barrel

Table 6.19 We still have a few moderate possible causes for our imbalanced wine

Process phase	Possible imbalance causes	Scenario 2 high and moderate likelihoods
Viticulture	Grape growing conditions and or varietal mismatch	High
Fermentation	Air exposure and bacteria	Low
Aging	Too little or too long in wood barrel	Moderate
Aging	Air exposure	Low to moderate
Aging	Used barrel cleaned but not sanitized	Low to moderate

Fig. 6.10 With our vegetal-fruit imbalance, we will investigate three materials: grapes, yeast, and ML bacteria

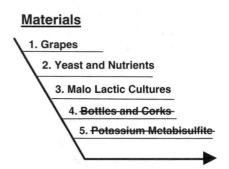

Materials

1. Grapes
2. Yeast and Nutrients
3. Malo Lactic Cultures
4. ~~Bottles and Corks~~
5. ~~Potassium Metabisulfite~~

6.4.1   Grape Materials and Imbalance

There are many hundreds of compounds in a wine grape and it may seem like there are an infinite number of flavors and imbalance possibilities. However, we have just begun this noble quest so we are going to give ourselves a fighting chance by considering a manageable number of groups. Start a list of possibilities. Wine is mostly water, alcohol, sugar, acids, volatile compounds, and other trace elements. Table 6.20 comments about the likelihood of possible material causes of wine imbalance. Let us investigate a few common material imbalances that may show up in our grapes.

One of our primary messages is always grow or select the best grapes you can. Pick grapes that are appropriate for your style and are healthy, ripe, and not damaged. The sugar acid imbalance can be driven during by the vineyard terroir, the grape varietal, the growing season climate, and the time of harvest.

Even if we do not get this optimally right, a sugar acid imbalance is not likely to drive our concern of fruit to vegetal imbalance. Table 6.21 lists sensory and chemical tests that can help reveal possible causes of our imbalance. Too much sugar with too little acid may indicate grapes were harvested late and have little acid backbone. Our scenario problem description indicates an imbalance of heavy vegetal offset by too little fruit. The most common material component of Cabernet Sauvignon that produces green pepper, or herbaceous, or vegetal flavors is a pyrazine, in particular,

Table 6.20 Typical grape material component imbalances potential causes

Possible imbalance causes	Likelihood	Comment
Too much sugar, too little acid	Low	High sugar/low acid can come from late harvests. High sugar fermented to high alcohol, but insufficient acid so tastes flabby or watery
Acid type imbalance	Moderate	Acidic red wines contain malic, tartaric, citrus, and other acids. A high acid type may produce a sour note, but not likely mask fruit with vegetal flavors
High relative concentration of methoxypyrazine (bell pepper taste)[a]	High	Methoxypyrazine is associated with bell pepper vegetal flavors in Cabernet Sauvignon
Excessive bacteria on skin	Low	Excess bacteria material may drive major flaws such as acetic acid and volatile acidity. However, not likely result in veggie imbalanced wine. This assumes use of sulfites, cleaning, and sanitization in the winery

[a]Refer to grape growing and harvest process sections for more discussion

Table 6.21 Sensory and lab tests provide great data to better understand wine imbalance

Possible imbalance causes	Tests	Comment
Too much sugar, too little acid	Visually inspect and taste grapes in vineyard Note skin, pulp, and seed characteristics Test sugar and acid levels	Tests help quantify imbalance levels of sugar acid concentrations. Review historical trends and associate good tasting wines to establish good and bad imbalance ratios
Acid type imbalance	Test acid types at harvest. Taste grapes and note skin, pulp, and seed characteristics	Quantified relative acid concentration levels provide objective imbalance characteristics
High relative concentration of methoxypyrazine (bell pepper taste)	Taste grapes and note skin, pulp, and seed characteristics. Chemical tests to isolate grape components	Methoxypyrazine is notoriously heavy in Cabernet Sauvignon driving bell pepper vegetal flavors
Excessive bacteria on skin	Visual inspection for healthy grapes at harvest. Screen for undamaged fruit incoming to winery operations	Expect some level of bacteria on vineyard grapes, but do not exacerbate by picking unhealthy or damaged fruit

methoxypyrazines (MP). We will be sensitive to vegetal bell pepper flavors going with Cabernet Sauvignon grapes.

Look to variety selection and viticulture practice for helpful preventive measures, since MP concentration has been shown to highly correlate with grape variety and grape growing conditions. Table 6.22 shows us a few preventive measures that can limit imbalance due to our winemaking materials.

Table 6.22 Great wine materials are truly established by healthy and ripe grape preventive measures

Possible imbalance causes	Preventive measures	Comment
Too much sugar, too little acid	Test before harvest, track balance of sugar and acid, and pick accordingly. Cannot add sugar, but possible to add tartaric acid if required	Test data provides best knowledge of when to pick. Nature and circumstances sometimes interfere, but get as close as you can. Note that you can add appropriate levels of tartaric acid
Acid type imbalance	Choose appropriate grape varietal for style of wine. Measure acid components and add to raise deficient levels. Optimize viticulture management practice.	Select best grape variety for your preference and growing region. Once chosen, adopt best vineyard practice, canopy management, harvest decision. Acid adjustments in the winery can help
High relative concentration of methoxypyrazine (bell pepper taste)	When using Cabernet Sauvignon grapes, provide best growing conditions. Perform winter pruning and canopy management during spring and summer, and fall	Methoxypyrazine is associated with bell pepper vegetal flavors in Cabernet Sauvignon
Excessive bacteria on skin	Visual inspection for healthy grapes at harvest. Screen for undamaged fruit incoming to winery operations. Use of sulfites in wine production will kill unwanted bacteria	Refer to basic winemaking chapter and volatile acidity chapter for handling bacteria

6.4.2   Balancing Three Important Materials in Finished Wine: Alcohol, Acid, and Tannins

Figure 6.11 is that classic seesaw image of balance. It not only represents balance, it also conveys the precarious situation in which conditions can change rapidly and one taste quality might shoot up and another falls down. Let us take a focused look at three critical components that make, or break, quality wine. These are alcohol, acid, and tannins. Grapes bring sugar to the fermentation process that creates the alcohol and we will talk about yeast next. Grapes also bring the acid and some tannin, but no alcohol as of yet. But we want to consider all three critical finished wine components; alcohol, acid, and tannins are front and center as we progress through making balanced wine. Let us take a qualitative look at how they work together. Do they complement each other and push the balance in the same direction? A balanced red wine includes all three and has the proper balanced levels of each. This discussion is based upon the exquisite thoughts and ideas expressed by Emile Peynaud (Peynaud & Blouin, 1983).

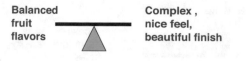

Balanced red wine when supple and not too hard. Can have good flavor, but needs good mouthfeel, ...

Fig. 6.11 A balanced red wine does not just have great flavor, it is supple, has good mouthfeel, and a lasting finish

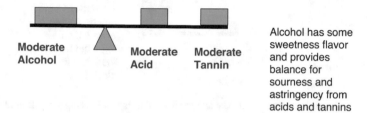

Alcohol has some sweetness flavor and provides balance for sourness and astringency from acids and tannins

Fig. 6.12 A quality red wine contains the right amount of alcohol to balance or offset the acid and tannin levels, each providing their unique and important flavor and sensory qualities

Wines benefit from the wonderful acids and tannins that come from the grapes, their skins, and aging in wood. The acids provide a crispness and freshness. The tannins allow longer aging in wood and bottle to provide well-balanced and complex aromas and tastes. Figure 6.12 illustrates a proper balance is enabled by the right levels of each. They provide that certain astringency or mouthfeel makes for a balanced enjoyable alcoholic drink. It is that very same alcohol that allows for higher levels of acids and tannins to coexist in a quality balanced wine.

Figure 6.13 shows high alcohol can be balanced with high acid and moderate tannin or as shown in Fig. 6.14, high alcohol can be offset with high tannin and low acid. There is a wide range of relative ratios of acid and tannin in grape varietals and aging methods. It is difficult to perfectly "dial in" the quantities you want but it is a good quality goal for better balance. However, the presence of alcohol allows for balance when your acid may be very high and your tannin moderate or low. Similarly, a higher alcohol quantity can balance out a higher tannin level in combination of a low acid. This gives the winemaker greater flexibility to craft a balanced wine.

Unfortunately, one can still easily end up with an unbalanced wine if the winemaker is not careful. With both high acid and high tannins, as shown in Fig. 6.15, it is likely you will produce a hard or very astringent wine. Of course, with too much alcohol, as shown in Fig. 6.16, the wine will be too boozy or hot and present as very unbalanced.

We have discussed grape materials and the offsetting synergies between alcohol, acid, and tannins.

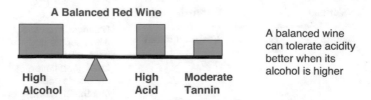

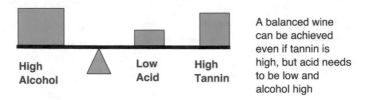

Fig. 6.13 A balanced wine can tolerate higher acidity when its alcohol is also higher

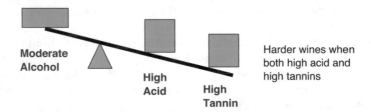

Fig. 6.14 A balanced red wine can also be achieved, even if its tannin is high, but its acid needs to be low and its alcohol high

Fig. 6.15 An unbalanced hard red wine can occur when both its acid and tannins are high and its alcohol is low or moderate

Let us take a deeper look at the output material of fermentation and material products from various yeast types. Yes, the primary process function is to convert sugar to alcohol, but there is more!

6.4.2.1   Malolactic (ML) Secondary Fermentation and Balance

MLF is a very common practice amongst winemakers to reduce the acid balance in many high-quality wines, particularly for red wines. Refer to Chap. 5.4.3 for background and discussion of RCA regarding possibilities of introducing excess VA.

If you prefer the lower acid for your style of your wine, then ML fermentation is a viable process. To properly start and finish ML fermentation, there are constraints to be addressed and conditions to be satisfied as shown by the inward facing arrows in Fig. 6.17. Use the balance guidelines to help decide if you actually want to lower the acid. If your wine is high in alcohol, you may not want ML. Once you have decided to implement ML fermentation, the risk is completing it successfully. Refer

An Unbalanced Red Wine

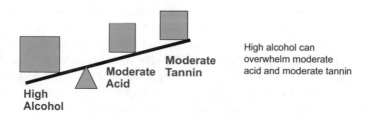

Fig. 6.16 An unbalanced red wine can be hot and boozy with moderate acid and tannins if the alcohol is too high

Fig. 6.17 ML lowers the acid level and changes the balance

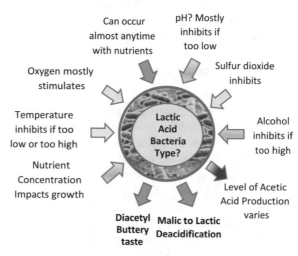

to Chap. 5.4.3 for a table of possible risks against a successful process, tests that can aid in diagnosing problems, and preventive measures that will help ensure ML start and finish properly. Wild ML bacteria strains can produce off flavors and acetate, and poor sanitation can harbor these bacteria in the winery. But since we are using commercial bacteria and address good sanitation processes in the Processes section, these are not significant. The only valid concern for ML bacteria to be a possible root cause for this scenario of low fruit to high vegetal characters is that ML bacteria can somewhat diminish fruit aromas, but this tends to be in delicate red wines like Pinot Noir.

6.4.3   Imbalance and Yeast Type

Could our imbalance have been introduced or enhanced by the yeast type? Refer to "The Principles and Practices of Winemaking" (Boulton, Singleton, Bisson, & Kunkee, 2010) for a comprehensive discussion of yeast in winemaking. In this section, we are not going to provide a comprehensive yeast overview, but illustrate our diagnostic and problem-solving approach, by focusing on our problem statement and the yeast type and grape varietal for this scenario. Our wine is made from Cabernet Sauvignon grapes and we used BM4x4 yeast.

The manufacturer advertises that this strain has been isolated by a research program in Tuscany Italy and claims it to be appropriate to both red and white wines. They also claim this yeast blend has the capacity of releasing a significant quantity of polyphenol-reactive polysaccharides into the fermenting must. What does this mean? What does it mean in terms of balance? It also has stated the following properties:

- Temperature range of fermentation: 16–28 °C.
- Moderate fermentation speed.
- Alcohol tolerance: 16%.
- High nutritional requirements.

We did not have a slow or stuck fermentation, and it appears our log book shows we kept the temperature within range, did not have too much sugar nor subsequent alcohol, and provided the necessary nutrients. Compliance with the manufacturer's specifications gives us higher confidence that we will have a normal and healthy fermentation. We also did not have any major glaring faults; however, we did have a subtler yet discernable imbalance.

What is the impact of polyphenol-reactive polysaccharides? Ugh, this could quite literally be a mouthful. We are not going to get into the detail chemistry; however, one can research the literature cited and uncover underlying properties. It is a complex topic. It also can be confusing to itemize impacts from polysaccharides as if they are one group with only one set of characteristics. It depends on their composition, their origin, and your particular grape juice. They might cause many different kinds of changes. They can change the color, the flavor, and or the mouthfeel. They can improve tartrate stability. They can change the mellowness or change the astringency sensation or impact haziness.

In our case, some may have come from our grapes and some may have come from the fermentation process using this BM4x4 yeast. The polysaccharides from yeast are mostly mannoproteins. There is a strong possibility that there will be a positive effect that will improve stability, mouthfeel, and the body of the wine.

However, even given the above, we do not consider it moderately nor highly likely that this yeast will significantly change the fruit to vegetal sensory balance. We could say it is possible, but not likely. We were able to note balanced prior vintages, Cabernet Sauvignon grapes, different growing region, same BM4x4 yeast, and fermentation conditions within range of manufacturers specifications. There is

Table 6.23 This scenario's imbalance due to grape material possibilities is likely because of a high concentration of methoxypyrazines, challenging growing site, and marginal vineyard practice

Possible imbalance causes	Likelihood for this scenario	Comment
Berry pulp material concentration to skin material concentration imbalance	Moderate	Our scenario is growing Cabernet Sauvignon in marginal region for this varietal. Drip irrigated too much. Failed adequate deficit irrigation
High relative concentration of methoxypyrazine	High	Given marginal terroir, less irrigation was critical. More frequent and more precise canopy management was needed to maximize available sun in cool climate

Fig. 6.18 Equipment can play a role in balance

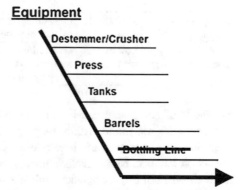

a small possibility that yeast material can cause imbalance, but it is not likely in this scenario.

Table 6.23 summarizes our current moderate or high possible material causes of imbalance in our wine. As you can see, there is overlap or tightly intertwined characteristics that tie many aspects of our situation together. It is hard to definitively isolate each, but we improve our chances if we can get the more probable suspects visibly high on our list. Let us move on and consider possible causes of our imbalance due to the equipment we used.

6.5   Equipment and Imbalanced Red Wine

We have shown the equipment fishbone of winemaking in Fig. 6.18 and crossed out the bottling line for this scenario because we discovered the imbalance after aging but before bottling. Equipment is necessary and helpful in making quality wine. However, each piece of equipment does have a chance of influencing ultimate wine balance.

Let us begin our root cause investigation with identifying the likelihoods of each equipment type on impacting wine balance. Table 6.24 comments on the general

Table 6.24 Each piece of equipment can impact balance, but with a different likelihood

Imbalance by equipment	Likelihood	Comment
Harvesting equipment	Low to moderate	Not a likely cause unless excessive leaves and debris are collected and not sorted out before crush
Destemmer/ crusher	Low	Need to check for broken crush equipment which might do more than break grape skin
Fermentation tank	Low	Refer to process and environment chapters for more on control of temperature and limiting oxygen exposure. Do not expect tank to be root cause or major contributor to imbalance
Press	Moderate	High pressure and or long durations of press might induce more bitterness or astringency from compounds in skins and seeds
Aging tank or barrel	Moderate to high	Long duration contact with tank or barrel material can impact balance of wine
Filtering and botting equipment	Low to moderate	Direct filtering of particles or microbes or fining of compounds can impact chemistry of wine and balance, stripping flavors

likelihood of equipment causing our imbalance problem. Grape flavor and aroma characteristics are found within the pulp, skins, and seeds. Equipment, in normal operation, that does not vary these relative concentrations, will not provide an early impact on balance. Most machine harvesting equipment or handpicked harvesting methods will not impact balance, unless excessive leaves and other material other than grapes get collected and are not sorted out before crush.

The fermentation tank holds the grape juice and yeast for a few days, while the fermentation process is underway. Refer to the process and environmental chapters for more discussion of related temperature and oxygen impacts.[4] This is typically a stainless-steel tank that will not impart balance chances to the process. The primary action is coming from the yeast and grape fermentation process itself. The fermentation tank is also a low likelihood contributor to wine imbalance.

The press, by the very nature of its primary function, is a possible cause of imbalance. We use a small hydraulic press, shown in Fig. 6.19, that is driven by standard water pressures around 35 psi. The press has an interior bladder that is inflated by filling with the water. Without any pressure the "free run" wine flows out. In the low to medium pressure range, more wine or alcoholic juice is pushed out, and the perforated screen catches the pressed skins and residue. At higher pressures, even more juice is pushed out, including other compounds from the skins and seeds. These are tannins and compounds that will increase the bitterness and astringency of the wine. This changes the balance of your wine.

---

[4] Refer to the Sect. 5.8 on environmental effects for additional discussions of fermentation temperature and oxygen impacts.

Fig. 6.19 The press will likely impact balance. With no pressure, this "free run" wine can be the fruitiest

The aging barrel material will surely impact flavor and aroma compounds that will change the balance of your wine. Refer to Sect. 4.6 regarding the process of aging and more on oak barrels and balance impact.

One important attribute of the barrel is size. Figure 6.20 shows a half barrel on the left and full barrel on the right. The size and dimensions will determine surface contact area with respect to volume of wine. The higher the ratio of an oak barrel surface area to wine volume, the more relative oak contact area. This can enable faster and more intense impact of oak characteristics into the wine. Let us compare a typical half barrel (30 gallons) with a full barrel (60 gallons). A simple approximation is assuming a cylindrical shape. More oak surface area for a given volume of wine produces relatively more oak flavor and aroma impact. For typical diameters and lengths, a half barrel has almost a 30% increase in surface area to volume ratio over a full barrel. This can change the rate and/or intensity of wine balance impact during the aging process.

A fresh new oak barrel also imparts more oak compounds as compared to older used barrels. Extended contact with wine during the aging process extracts more and more compounds from the barrel. At some point, the used oak barrel has little discernible elements left. The oak barrel becomes neutral or has no effect on imparting flavors and aromas. This is a "perishable" piece of equipment. Your winemaker

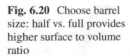

Fig. 6.20 Choose barrel size: half vs. full provides higher surface to volume ratio

style choices and pocket book will guide you on how many times to reuse an oak barrel.

Table 6.25 has dropped the low to moderate probability possibilities and retained the moderate to high ones. It lists tests that will aid in our investigation and identifies contributing factors. The essential press does bring imbalance side effects. As noted in other sections, excess of pressure can extract unwanted high levels of astringent and bitter compounds from skins and seeds. For aging, the barrel or tank material type, size, contact area, and volume all play into duration of aging and degree of balance.

Preventive measures, listed in Table 6.26, will improve your chances of achieving the wine style and balance you desire. By disciplined and rigorous application of preventive measures, one can reduce the number of possible imbalance risks. In addition, having a good set of winemaking skills with appropriate preventive measures ensures a good baseline to check against. This can highlight shortfalls and contrast events that may have caused your wine imbalance. You will notice an overlap in preventive measures in the various sections. This is intended to provide a more complete and synergistic approach that leaves no gaps or errors behind.

It is good practice to fill a new barrel with water and let it soak for 24 hours. This will help the wood expand and seal and prevent wine leaking out or air coming in.

Table 6.25 Include tests along with equipment use, particularly for those moderate to high imbalance contributors

Equipment possibly contributing to imbalance	Tests to understand equipment use for moderate to high possibilities
Press	• Monitor and inspect to collect all free run wine • Monitor press pressure • Refer to other process and people sections for related tests
Aging barrel or tank	• Note barrel size and volume • Update history of number and time of barrel use • Frequent inspection of bung seal • Monitor barrel room temperature • Refer to other process, material, and people, environment sections for related tests

Table 6.26 Preventive measures applied to use of your equipment can aid in improving your wine balance

Equipment possibly contributing to imbalance	Equipment use preventive measures for moderate to high possibilities
Press	• Clean before and after use • Collect all free run wine • Adjust press pressure to lowest setting if possible • Do not conduct second press • Refer to other process and people sections for related tests
Aging barrel or tank	• Clean before and after use • Choose barrel material, toast, size, and adjust aging time to fit style • Fill with water to swell wood 24 h before use • Frequent inspection of bung seal • Top-off frequently to limit air • Keep barrel aging room temperature <70 °F • Lengthen barrel aging time if want more oak, vanilla, smoky style • Shorten oak barrel aging duration or use stainless if want more fruit forward style • Refer to other process, material, and people, environment sections for related tests

It is critical to frequently top-off barrels during aging to prevent the wine from oxidizing with too much air in the headspace.

In our current scenario, we have checked our records and understand that we did a reasonable job with our press and associated preventive measures. We collected all of the free run wine, kept pressure relatively low, and did not conduct a second press. However, we did use oak barrels and noted that there was some possible missed top-off events and that we did age in the barrels a long time. At this point in the winemaking process, we cannot raise the fruit level, but we might mask it with other oak compounds or air exposure. This would further reduce the fruit to vegetal balance.

Table 6.27 shows the summary of only one item left on the equipment fishbone for this scenario. This is aging in an oak barrel with insufficient topping off causing imbalance in our wine. Our equipment is absolutely necessary for making quality balanced wine, however, if we are not careful, we can introduce further risks to cause imbalance. Let us now take a look at the operators of our winemaking equipment, our people. Are there possibilities, that they may have unknowingly or inadvertently introduced imbalance into our wine.

6.6   People and Wine Imbalance

Nature provides us with the wonderful grapes and yeast to make fantastic well-balanced wine. The people on our team perform the winemaking tasks and further enable the wonderful balanced finished wine. We are not going cover "imbalanced" people in this book; however, we will take a stab at wine imbalance and the effect otherwise normal people might have on the process. This material expands upon ideas discussed in the Sect. 5.5 in the volatile acidity scenario regarding people effects. Figure 6.21 is the people fishbone showing possible problems originating with the vineyard, winery, or laboratory staff. Each presents our team with the risk of causing or exacerbating the imbalance in our wine.

Table 6.27 For this scenario, we have narrowed down the equipment chances of wine imbalance to our oak barrels and their use

Equipment cause of wine imbalance this scenario	Comment
Aging barrel or tank	• Moderate to high possibility due to long duration and possible air leaks in oak barrels • These conditions during a relatively long oak barrel aging could mask already weak fruit flavors and aromas

Fig. 6.21 Wine imbalance may be introduced by any member of the team. Vineyard maintenance, wine production, and laboratory operations

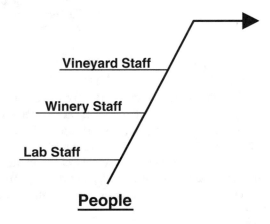

We encourage specialized training to avoid wine imbalance and enhance wine balance. The key is to pay careful attention to competing forces that may lead to imbalance. For example, when you test for sugar, do not forget to test for acid. When you test and note you have high alcohol, do not forget to test and take steps to also keep the acids high and tannins moderate to high as well. Some of these steps can might be implemented in subsequent aging or may have to wait until next vintage.

Training and instruction are a good place to start. The vineyard requires attention all year around—not so much during the dormant winter months, but quite a bit during the spring, summer, and fall. Vines will probably grow if left alone, however, getting grape quality and yield requires vineyard management.

For this scenario, we would like to grow Cabernet Sauvignon grapes in the Santa Cruz Mountains and in a canyon no less. This is our scenario's limited sun and varying temperature circumstance. This drives the need for planning and sustained high level of vineyard care.

The winery staff takes the beautiful grapes from the vineyard and produces the finished wine. The production crew in turn relies upon the test results and periodic wine assessments to guide the course of action and details of each production step. The people and actions are all interconnected and important to make it to the finish line with a balanced wine. Table 6.28 comments on the likelihoods of our imbalanced wine problem being caused by the various staff.

We have discussed picking the right growing region for the variety of grape you prefer; however, many real-world constraints may drive this away from the best match. This can be mitigated and overcome with good viticulture practice which needs good planning, sustained vineyard management, and good training of your viticulture crew. If your grapes are struggling to ripen before the season ends, there may be an inherent vegetal to fruity balance in your grapes. Table 6.29 identifies activities to aid in identifying people-related causes of our imbalanced wine.

Planning, training, executing proper soil treatments, deficit irrigation, canopy management, and smart harvest decisions are all needed. A shortfall in any one of these may raise the risk of imbalance too high. A vigilant well-informed vineyard

Table 6.28 Your winemaking crew can enable great balanced wine

Potential risk to imbalance by the crew	Likelihood	Comment
Vineyard	Moderate but low with good preventive measures	Low with good preventive measures including planning, growing season maintenance, and smart harvest decision
Winery operation and production	Moderate but low with good training	Establish good winemaking plan and provide good training
Laboratory staff	Moderate but low with good training and procedures	Perform appropriate tests, interpret the results, and apply appropriate actions and or corrections. Keep record to apply to next year's vintage

However, there are risks with an untrained crew

Table 6.29 Your winemaking team will help with tests and observations

Potential risk to imbalance by the crew	Activities to aid in isolating people related Imbalance contributors	Comment
People related errors in the vineyard	• Audit plans, instruction and training • Frequent vineyard inspections, maintenance, and feedback on status • Ask vineyard crew to go beyond maintenance, voice observations and recommend areas to improve	Collect frequent observations of vineyard growth. Encourage everyone on the team to provide feedback on grape vine conditions
Winery production team errors	• Audit plans, instruction and training • Observe operations that sort fruit coming into the winery • Production crew should maintain "juice to wine" logbooks, audit results, and analyze trends	Develop good plans, but also track logbooks to find possible errors or patterns. We had good intentions to top-off wine frequently, however, winery crew distractions and external events caused a number to be missed
Laboratory staff errors	• Audit plans, instructions, and training • Audit logbooks of "juice to wine" treatments	Provide training and periodic updates to maintain skills and keep up with latest developments

Good feedback from your team helps prevent imbalance

crew can overcome these challenges by starting the day with a review of the planned processes. Winery and laboratory operations can also enhance imbalance risks that are already marginal from the grapes. A misstep by our vineyard, winery, or laboratory operations crew can further raise the likelihood of imbalanced wine.

Proper training of your crews is very important to sustainable quality of your wine. Training needs to span use of equipment during the day and nighttime use. Operations may extend past sunset. Evaluating, updating, and correcting shortfalls in your training efforts can be enhanced with frequent checking and auditing of your training plans and results. On-site vineyard checks, inspection of vine and grape health, and viticulture performance, and actual crew maintenance provide insight. Similarly, periodic sampling of wine quality and production crew performance keeps the quality up. An informed and up-to-date laboratory crew keeps the wine quality high.

The U.C. Davis Viticulture and Enology Department and other reputable organizations provide frequent education and training courses in every aspect of grape growing and winemaking.

Training is the cornerstone of building a quality winemaking crew and high-quality product. This scenario's problem is imbalanced wine with a heavy vegetal to fruit characteristic. Training programs themselves should consist of high quality, clear, and thorough instructions. They can also emphasize maintaining balance and reducing risk of imbalance. This ensures instructions and testing of complementary balance characteristics.

We discussed material combinations of sugar, pH, and acid in the grapes and sensory combinations of alcohol, tannins, and acids in the wine. Ensuring your training plans and trained crews are keeping track of these leads to higher quality balanced wines. Cross-training laboratory staff on tests methods provides for additional flexibility for coverage in peak high volume or time critical needs.

Our Cabernet Sauvignon may not have the best terroir in this scenario. However, these limitations can be overcome with thoughtful care and training of the vineyard, production, and laboratory crew. We believe our training materials and instructions were not fully up to the task and have assigned them a moderate likelihood of causing our imbalance in the people bone.

The key to thoroughly evaluating the people bone of the fishbone diagram is to look for human error. If we have a good procedure for leaf removal in the vineyard but the summer intern was not trained to follow it, then the root cause is in the People category (vs. the Process category.) If we have a good process for barrel topping and all the winery crew is well trained but the owner or someone from the tasting room does frequent barrel tastings without telling the winery staff, then the root cause is in the People category. So, it is important when evaluating this bone to interview or audit everyone and look for inconsistencies or anomalies. Table 6.30 summarizes comments on actions and preventive measures you can take to better train and inform your staff to minimize people related causes of imbalanced wine.

You do not want this to be punitive, you want everyone to be forthcoming with their answers. Ask each person to describe how they perform a task and look for any outliers. One person may use a different piece of equipment or use a different measurement device, leading you to a different root cause category. Or you may find that one person (and not necessarily the "new guy") does not really follow the same process as everyone else and that difference can in fact cause the problem you are trying to solve.

Table 6.31 identifies the most likely people related training shortfall that could be the cause or probable contributor to our imbalanced wine problem.

6.7 Measurements and Wine Imbalance

Before we dive into more root cause analysis, we are going to take a step back and review some basics. Figure 6.22 is our tailored fishbone group for measurements.

Let us discuss measurements and errors to uncover and prevent unbalanced wine. Measuring grapes in the vineyard is important to a balanced harvest decision; sugar, acid, and pH. Measuring grape juice in the winery is important to establish a good early start on balanced winemaking. Measuring sugar, acid, pH, alcohol, and tannins in the winery is also important to maintain balanced wine.

Table 6.30 Preventive measures rigorously applied will improve your chances of a balanced finished wine

Potential risk to imbalance by the crew	Preventive measures for team and wine imbalance	Comment
People related errors in the vineyard	• Training in canopy management tailored to region and terroir • Training in balanced harvest decisions, i.e. sugar high enough at same time acid not too low • Vineyard manager keep logbooks throughout growing season • Recognize team accomplishments	Canopy management appropriate to terroir is critical. Balanced plan for harvest decision. Recognize team with harvest party (good wine balanced with good food!)
Winery production team errors	• Training in production, juice to wine, end to end. Highlight tracking balance of parameters • Keep logbooks or travelers and highlight competing balance parameters • Recognize accomplishments	Training should also include cross train in different tasks to minimize repetitive task errors and keep enthusiasm high. Refer back to prior vintages for historical comparisons
Laboratory staff errors	• Train and retrain with emphasis on testing offsetting balance parameters • Train in sensory and chemistry measurements • Jointly audit measurement results, note trends, flag imbalanced pairs, and highlight outliers	Do not just test for offsetting wine balance parameters such as alcohol, acid, and tannins. Review, analyze, and compare current and historical trends. Work to connect groups of characteristics and trends to finished wine

Table 6.31 Given challenging growing location for Cabernet Sauvignon, we consider vineyard training materials moderate likelihood for cause of people related imbalance problem

Potential risk to imbalance by the crew	Likelihood	Comment
Vineyard management	Moderate given growing region not optimal for Cabernet Sauvignon and canopy management shortfalls	Vineyard management might overcome challenges of this growing region. Training of vineyard crew is critical element. The training program, instruction materials, and frequency of training must be high quality and truly balanced with the high quality of the wine itself

Fig. 6.22 Fishbone subgroup for measurements. Note we have crossed out weights and volume measurements as unlikely given our good practice in these areas for this scenario

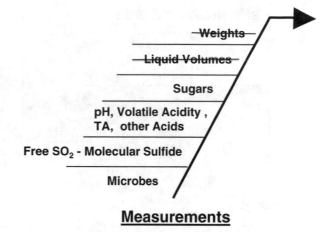

Measurements

Figure 6.23 is graphically depicting three measurement groups. The first is an early measurement series of the grapes in the vineyard. These are primarily of sugar, acids, and pH. The second group are measurements of grape juice in the winery. The winery measurements are later but are also of sugar, acids, and pH. This allows tracking and trending. The third group are also measurements of sugar, acids, and pH; however, we have added measurement of tannins and alcohol. This comes after fermentation and allows further tracking and trending. The overlap in components allows you to follow how your wine is always changing and evolving. Remember, measurements can be chemical concentration in nature and should include sensory as well. One of the messages is that early measurements may have large uncertainties with respect to where the finished wine is headed; however, they also provide powerful objective indicators of how to tailor your winemaking methods to achieve the style you want and maintain the terroir of your vineyard.

Refer back to the Sect. 5.7 discussion of measurement errors associated with weights, volumes, sugars, acids, and pH. We are going to address in this chapter a discussion of alcohol and tannin measurement errors. Then we will move onto possible imbalance risks of various measurement errors and how to prevent them. We consider sulfites important to controlling spoilage mechanisms in wine and should add more or less to maintain the quality, but not change the balance or style of wine.

Making measurements early and throughout winemaking helps provide good knowledge of wine condition status, forecasting future possibilities, and guiding subsequent steps. Sugar can be measured in vineyard and at harvest and can forecast alcohol content in finished wine. The error may be relatively large at the time, but it will guide winemaking steps for a better-balanced wine. Let us review alcohol and tannin measurements, before we consider these three groups of measurements.

Home winemakers or small wineries may not want to spend the resources for sophisticated measurement equipment. Winemakers should still understand the errors or uncertainties in the processes they chose to follow. Understanding the cause of this scenario's imbalanced wine may depend upon it.

Early measurements

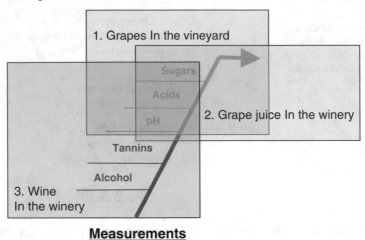

Measurements

Fig. 6.23 Early measurements may have larger errors, but early forecasts lower future risks

We are going to focus on errors from three alcohol measurement approaches shown in Table 6.32. The first is extrapolating from sugar at harvest, the second is a more traditional Distillation and Hydrometry approach, and the third is Ebulliometry. Reference Iland (Iland, Bruer, Edwards, Weeks, & Wilkes, 2004) for methods and procedures.

Measuring grape sugar content is a long time standard critical aid in determining grape ripeness and guiding when to harvest. It can also aid in projecting the final wine alcohol concentration. Figure 6.24 shows a histogram plot of % alcohol predictions based upon six nominal brix levels at harvest. For each of the brix levels, there are three columns showing a minimum, nominal, and maximum alcohol level using a nominal ± one-degree brix uncertainty in the sugar measurement and the rough scaling rule of 0.55 times brix equals alcohol concentration. The accuracy is not enough for final labeling laws; however, it is a reasonable guide for maceration, fermentation, and aging methods to avoid an imbalanced wine. The alcohol extrapolation figure shows errors can be large, however, they may still be limited enough to guide your next step or two in winemaking.

As we saw earlier in Sect. 5.4.2, we may want to guide the process to increase the level of acid and or tannins if we are high in alcohol, or vice versa. We will still need to make a more accurate alcohol measurement for labeling laws; however, some knowledge early can help with better balance in subsequent winemaking steps. If you can afford resources and complications of the Distillation and Hydrometry, or Ebulliometry, or other spectroscopic variants to get 10× the accuracy in alcohol levels early in winemaking, it can help make more informed decisions.

Table 6.33 comments on the likelihood of measurement errors causing our imbalanced wine. Note that we consider the measurement errors to be low, with only

Table 6.32 Alcohol measurement error contributors

Alcohol measurements	Typical error magnitudes	Comments on errors
Extrapolating to final alcohol from sugar at harvest	Assume grape sugar is within ±0.5%. Scale factor from grape sugar to finished wine alcohol can be between 0.5 and 0.6, alcohol can range over ±1%, reference Fig. 6.24	Grape sugar errors can come from poor representative sampling of vineyard, temperature variation between measurements, refractometer measurement error, or hydrometer error of crushed grape juice yeast fermented grape juice produces between 0.5 and 0.6 times the sugar content to alcohol
Distillation and hydrometry	Ethanol by volume accuracy: ± 0.1%	Volume errors of wine or distillate, cleanliness of equipment, or leaky seals can contribute to distillation errors. Hydrometry without proper temperature correction, properly sized and cleanliness of containers can also contribute to hydrometry errors
Ebulliometry	Ethanol by volume accuracy: ± 0.1%	Incorrect temperature of boiling points or inefficient operation of reflux condenser can contribute to ebulliometry errors

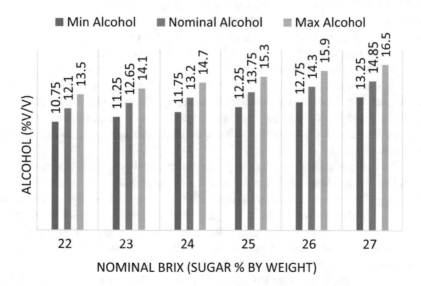

Fig. 6.24 Errors in brix measurement and extrapolation make for larger error; however, some early knowledge can guide winemaking. Grape sugar is not just for harvest decision

Table 6.33 Isolated measurement errors likely small, however, combined measurement errors could be a moderate contributor

Possible causes	Likelihood	Comment
Sugar concentration, acid concentration, and pH measurement errors in the vineyard	Low to moderate	There is some level of measurement error in the vineyard due to small sample sizes and the use of portable, less accurate instruments
Sugar concentration, acid concentration, and pH relative errors of grape juice in the winery	Low but not zero	Typical winery lab measurements of the grape juice are affordable, practical, and have sufficiently small errors to provide the foundation for good winemaking
Wine sugar, acid, pH, alcohol, and tannin relative measurement errors in the winery	Low but not zero	Typical winery lab measurements of the wine are also affordable and sufficiently accurate. A more likely issue is infrequent sampling or not at critical junctures, but these are process or people errors, and not specific measurement errors

slightly higher likelihood in the vineyard where sample sizes are small, portable equipment is less accurate, and personnel may not be trained chemists. Measurement errors are not likely to cause our imbalance problem.

Since wine is dynamic—changing with time—it is important to measure frequently enough to take the necessary actions associated with the collected data. If, for example, you follow a rule-of-thumb recipe for potassium metabisulfite additions but wait a year to test the free SO_2 level, you may find your wine to either be over-sulfided or not protected adequately against spoilage microbes.

Let us now discuss tannins and this scenario. In general, the phenolic content is responsible for color, mouthfeel, flavor, and aging potential in wines. The phenolic tannin concentrations vary significantly amongst red wine varieties, blends, and grape growing regions (Habertson et al., 2008). However, increased tannin concentration is well understood to cause increased level of astringency in wine. In addition, higher levels of tannins do allow wines to age longer. The source of tannins within the grape is predominantly in the skins and the seeds. So more or less tannins can be influenced for a given grape by spending more or less time on the skins and seeds.

Cabernet Sauvignon does seem to be a grape varietal type that is relatively high in tannins. The levels of tannins in finished red wine are also considered to be driven by the grape variety and the winemaking methods. Having knowledge or a measurement of the tannin concentration in your wine can help in deciding when to press or blend and make a balanced wine. This wide variation in tannin concentrations and ability to adjust time on the skins does indicate it is important to taste and test your own wine to make better decisions during the winemaking process.

If you are going to test for astringency, it has been shown that the tannin concentration does correlate well. There may be other contributors; however, tannin concentration is a good indicator of astringency. What concentration level is important? There are high astringent Cabernet Sauvignon wines with tannin levels around 800 mg/L or higher. There are also low astringent Merlot wines with tannin levels around 200 mg/L. Let us assume these are representative. We then want to have a test error that is small compared to these magnitudes. This roughly means test error in the 20 to 80 mg/L. Most tests such as protein precipitation assays provide better accuracy and precision. Most trained sensory tasters can distinguish between these low and high levels.

This scenario imbalance problem is too high vegetal to fruit. Measurement errors are small contributors to imbalance conditions in this case. As summarized in Table 6.34, we consider measurement error an unlikely root cause.

6.8 Environment and Wine Imbalance

We have entered the environmental bone of our RCA as shown in Fig. 6.25. Our scenario major problem is imbalance. Our scenario also does not use grapes grown nor facilities located in a high humidity environment. We also have established a relatively clean processing environment so that leaves the most significant driver as temperature and time. We are going to break the RCA assessment into three significant winemaking phases: the vineyard and harvest decision, fermentation, and aging. As a part of these winemaking process steps, we will look at sun exposure and temperature to illustrate our root cause analysis method.

Table 6.34 This scenario's imbalance is not likely due to measurement error

Possible causes	Likelihood	Comment
Wine sugar, acid, pH, alcohol, and tannin relative measurement errors in the winery	Unlikely	This scenario problem had a low fruit to high vegetal balance problem. Given the moderate ranges of other wine characteristics and measurement program, measurement error unlikely

Fig. 6.25 For this scenario, we will focus on sun exposure and temperature impact on wine balance

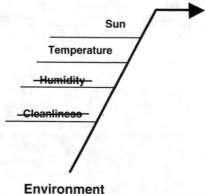

Red Wine Balance Environment

Sun

Temperature

Humidity

Cleanliness

Environment

The sun and solar radiation flux on the plants is arguably the most important environmental element enabling photosynthesis. Each vine and leaf has certain positions, orientations, and shading with respect to the sunlight. The growing region latitude, local geography, and hillside slopes all contribute to the hours of sun per day and radiation angles of solar flux impinging upon our precious vines. The vineyard orientation, vine spacing, and trellis design further define the daily solar flux. Vineyard canopy management provides one of many methods to optimize solar flux and shading throughout the growing season. The solar environment is highly likely to influence the growth of our berries and impact the balance of colors, aromas, and flavors.

In hot grape growing regions, ripening occurs very quickly and there is a reasonable risk that the grape acids may be too low. In cooler growing regions, there is a risk that the grapes will not ripen soon enough and that sugar and other important components will not accumulate to the necessary levels. Similar to the solar radiation issues discussed above along with a few other factors: latitude, geography, local terrain, proximity to water, vineyard orientation, trellis design, and canopy management will all play a role in vine and grape temperatures. These temperatures will also strongly influence grape ripening and component balance. As shown in Table 6.35, sun and temperature not only are big driving factors of growth and potential imbalance in the vineyard, they also have significant influence on the rates and quality of fermentation and aging.

Fermentation is one of the most important winemaking steps. We have discussed the process in the basic winemaking section and the prior VA problem scenario. This is where yeast converts sugar to ethanol, carbon dioxide, and heat. It has been well established that the yeast population grows exponentially with temperature, assuming there are corresponding nutrients to feed the yeast. At some point there are yeast killing effects that come into play, such as ever-increasing alcohol.

Aging is a long duration winemaking phase that is also very sensitive to temperature and time. So much can occur over periods of 12–24 months. It may casually

Table 6.35 Sun and temperature drive balance in the vineyard, fermentation, and aging

Process phases, sun, and temperature	Likelihood of balance impact	Comment
Sun and temperature in the vineyard	High	Sun provides the most important source of photosynthesis. Temperature over time drives metabolism of berries. It is a big driver. The balance of sugar to acid will go up significantly during a heatwave
Temperature/time of fermentation	High	The rate of yeast population growth and most reactions increase with temperature. Hotter or colder fermentations introduce many different characteristic end products
Temperature/time of aging	High	Hotter or colder long duration aging drives balance of many different flavor and aroma components. The temperature coefficient of many chemical reactions is about two, meaning at about 10 °C hotter, the reaction occurs at twice the rate

appear like the wine is in a sealed container, however, there is evaporation out, air coming in, and all sorts of wine components changing. Temperature and time make these many effects go faster or slower.

6.8.1 Temperature/Time Impact on Balance During Grape Growing and Harvest

We have discussed and provided references to basic viticulture practice in the book. Let us hypothetically assume that our vineyard provided sufficient sun, water, and nutrients for our grapevines and we implemented the corresponding appropriate canopy management. In this idealistic situation, let us consider two temperature conditions, one discernibly hotter than the other. The effect would likely mean the rate of ripening would accelerate in the hotter condition. Sugar would accumulate quicker and acid would correspondingly reduce earlier. We have created Fig. 6.26 to illustrate this which was inspired by the work and references in the Biology of the Grapevine (Mullins, Bouquet, & Williams, 1992). Note when sugar rises and acids fall approximately along an s-curve, the crossover "pseudo" balance point shifts significantly to the right when the temperature cools down. Accumulating historical temperature, sugar, and acid data from your grape source vineyard year after year can guide next year's harvest plans.

One of the takeaways from this figure is the sensitivity of grape component concentration to temperature and time. Not all components go up at the same time.

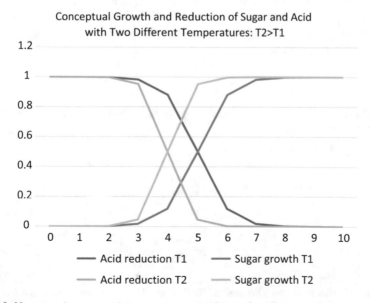

Fig. 6.26 Note rates increase with temperature and balance of sugar to acid changes

The balance between sugar and ripeness is rising and acid is falling. The crossover point is moving to the left and occurring early. The criteria for "balance" and harvesting with reasonably high brix and the right small amount of acid are also moving to the left and earlier. If it is hot and you are on the steep part of the curves, big balance changes can occur very fast. Stay close to the temperatures and conditions in your vineyard.

6.8.2 Temperature/Time Impact on Quality and Balance During Fermentation

We add yeast and yeast food to our grape juice to start fermentation and convert the sugar to alcohol. A temperature increase drives the yeast population up and the sugar to alcohol transformation increases. Yeast killing effects include the ever-increasing ethanol concentration.

This means there is probably an optimal temperature. Table 6.36 lists the min, max, and average temperature and duration of seven different vintages. The quality assessment is from various wine competition judges. For this example, we are also going to assume there was no preceding juice treatment like extended maceration. If this is the only time on the skins, the shorter time durations of this phase may have adverse effects on red wine. There is less color, flavor, and aroma extraction with shorter fermentation times. Temperature and time require a significant balance during the fermentation process.

Our RCA approach began by looking at our problem and effect statement. We have audited our winemaking records of past vintage fermentations and noted the different temperature and time characteristics. This data set is by no means a controlled experiment; however, it may provide some interesting clues and allow us to illustrate RCA techniques.

Table 6.36 Historical experience with different quality wines and corresponding different fermentation temperatures and time

Cabernet sauvignon fermentation					
Vintage[a]	T min	T max	T average	Duration	Quality
2010	64	73	69–70	13	High
2011	70	87	78–80	8	Low
2012	69	81	75–76	14	Moderate
2013	66	76	72–73	8	High
2014	67	82	77–78	6	Low
2015	71	83	78–79	6	Low
2016	62	76	72–73	8	Moderate

[a]Grapes are from different growing region: 2010 from Ukiah, 2011 and 2012 from Santa Cruz mountains vineyard A, and 2013, 2014, 2015, and 2016 from Santa Cruz mountain vineyard B

Table 6.37 These vintages use the same source grapes and same yeast

Cabernet Sauvignon fermentation					
Vintage	T min	T max	T average	Duration	Quality
2013	66	76	72–73	8	High
2014	67	82	77–78	6	Low
2015	71	83	78–79	6	Low
2016	62	76	72–73	8	Moderate

Note this pattern of temperatures, duration, and quality

It is very intriguing that 2010 and 2012 have longer time durations and relatively cooler temperatures and result in moderate to high quality. However, they have different grape sources from different growing regions and conditions. Because they are different grape source conditions, for this RCA and this focus area, we are going to eliminate them from this argument. Table 6.37 is the data for 2013–2016 for grapes all from the same vineyard. However, always keep in mind that multiple contributing factors and layers of preventive measures may all conspire to produce an imbalanced, or balanced, quality wine.

We have narrowed down to four historical vintages. This is a limited data set; however, for our grapes and fermentation practice, we can see a pattern of lower quality wine associated with shorter duration and higher temperatures. This in itself is not sufficient to isolate a true root cause; however, it is a strong indicator of a possible contributing factor. Part of our thorough RCA method does of course involve working through all the other bones and areas for finding additional contributing factors, weighing them, and building up a prioritized set of contributing factors and preventive measures to improve next vintage.

6.8.3 Temperature/Time Impact on Balance During Aging

Note that most winemaking chemical reactions have a temperature coefficient of around two which means that for a given temperature that is 10 °C warmer, the reaction occurs at about twice the rate. There are of course exceptions, like enzyme aided reactions, components added, or components deleted. However, for most part, it is about two. Controlling the temperature provides the opportunity to control wine reaction rates. Uncontrolled temperature introduces unknown risk. Cellars are generally cool and stable. Our RCA method includes testing, tracking, and understanding your own conditions. You can also compare with other known successful standards and common practice like those that have been published in surveys of California wine producers (Cooke & Berg, 1984).

6.8.4   Preventive Measures for Improving Balance of Fruit to Vegetal Characteristics Involving Temperature

We have presented background information showing sensitivity of grape and wine characteristics to temperature and time changes. Measuring and tracking temperature at frequent and critical winemaking process steps provide insight (refer to Sect. 5.8, for example). There are a number of preventive measures or strategies that principally reside in the vineyard, but also present in the winery, that can aid in improving fruit to vegetal balance.

6.8.5   Likely Environmental-Induced Causes for this Scenario's Imbalance

We have three winemaking phases that are highly sensitive to vineyard sun and temperature and winery operation temperatures. Table 6.38 comments on the various temperature or sun exposure aspects. For this imbalance scenario with a heavy vegetal to fruit imbalance, the table shows us moderate to high possibilities.

We have done a systematic investigation of possible causes. We have done a first cut at the likelihood of being the root cause or contributing factor. We have a prioritized list, but it is still fairly long. We have identified helpful tests and preventive measures; however, this is still a long list. Note that for this scenario, our region, the grape variety, the level of sun, and temperature are all elements of the high likelihood causes of our problem. Table 6.39 summarizes the high and moderate likelihood causes for our imbalanced wine.

Table 6.38 You can influence temperatures through trellis design, canopy management or directly with temperature controlled tanks and cellars

Sun and temperature preventive measures	Temperature control aids
Sun and temperature of vineyard	Orient vineyard rows north south, use VSP trellis design, and canopy management to maximize sun and temperatures. Frequent and smart canopy management to allow filtered sun, but limit sunburn during summer. Frequently measure and track temperature, forecasts, sugar, acid, and pH close to harvest
Temperature and time of fermentation	Measure and control fermentation tank just above min to activate yeast. Stay below higher temperatures for needless excessive rate of alcohol fermentation. Want long duration fermentation, preferably 1–2 weeks
Temperature and time of aging	Maintain aging cellar at cooler temperatures (<60 °F). Longer barrel aging may mask fruit with oak and vanilla characteristics. Long barrel aging may also enhance risk for excess oxygen exposure

Table 6.39 High and moderate likelihood imbalance causes identified

Possible imbalance causes	Comments	Likelihood
Viticulture	Grape growing conditions and/or varietal mismatch	High
Environment	Sun and temperature in vineyard	High
Material	High concentration of methoxypyrazine in Cabernet Sauvignon	High
People	Special canopy management with training needed for varietal and growing region	Moderate
Environment	Temperature/ time during fermentation	Moderate
Environment	Temperature/time during aging	Moderate
Aging	Too long in wood barrel masking light fruit	Moderate
Equipment	Aging in barrel risk of air leaks and masking already weak fruit characteristics	Moderate
Material	Pulp to skin material concentration too high	Moderate

Many possibilities indicate a lot of ground to cover

Table 6.40 We have three equal high likelihood causes

Possible imbalance causes	Comments	Likelihood
Viticulture	Grape growing conditions and/or varietal mismatch	High
Environment	Sun and temperature in vineyard	High
Material	High concentration of methoxypyrazine in Cabernet Sauvignon	High

Can we prioritize these?

Let us continue the quest by diving deeper and breakdown the three high likelihood causes.

6.9 Deep Dive: The Next Layer of Prioritization

We want more insight. We will rarely if ever have the resources, time, nor interest to implement preventive measures for everything. We will, however, want to attack the most likely possibilities. Let us break them out even further. Let us take a closer look at the three highly likely suspects: one in viticulture, one environmental, and one material item.

Our viticulture practice, our environment, and the material we used are each a highly likely suspect. Let us check for more clues in our problem statement, historical records, and current situation. Let us prioritize our three suspects within this high category in Table 6.40.

We are growing Cabernet Sauvignon grapes. This variety comes with a high concentration of methoxypyrazine. This dominant material component makes for a challenge for balanced grapes and balanced wines. We have placed trellis design, growing region soils, water (rain and or irrigation), and canopy management all into the viticulture category. We have separated out temperature and sun into a third. Let us define various tests, inspections, and assessments in Table 6.41 that will help us sort out the biggest driver.

Let us breakdown our hillside vineyard. Different parts have different sun, temperature, water, and or soil conditions. This might correlate with more fruit or more vegetal characteristics. Are the sun or shade conditions different for the north, south, east, or west sides of the vineyard? Are there steep hills or shade trees that limit hours of sun? Are we using a VSP trellising approach and do the rows run north south? What are the temperatures throughout the growing season? Are there good diurnal temperature swings from day to night? Figure 6.27 is a plot of high and low temperatures during the growing season up to veraison where the berries change color. Are the temperatures different at the top and the bottom of the hill? Is the irrigation even or different throughout the vineyard? Is the ground water and soil moisture different throughout the vineyard? Do we have quantitative data to analyze our situation?

Table 6.41 These tests and actions will help us figure out the biggest drivers, prioritize the "highs" and guide application of resources

Possible causes	Tests to help rank imbalance cause likelihoods	Comment
Viticulture: Region, climate, and canopy management	• Inspect vineyard canopy sun exposure and shading • Note weather, rain, and ground water • Assess soil, nutrients, drainage	Cabernet Sauvignon compounds and growing season may limit quality
Environment: Sun and temperature	• Assess vineyard latitude, orientation, and trellis design • Inspect sun hours and shading all season and different parts of the vineyard • Measure vineyard temperatures	Canopy management can aid in marginal levels
Material: High concentration of methoxypyrazine (MP) in Cabernet Sauvignon	• Given cabernet variety selection, greater insight and care in marginal growth regions is needed • Drives tests, inspections, and assessments of above environment and viticulture practice	High quality Cabernet Sauvignon is a challenge. Excessive vine growth, over-watering, and cluster shading exacerbates MP. Pay greater attention to canopy management details

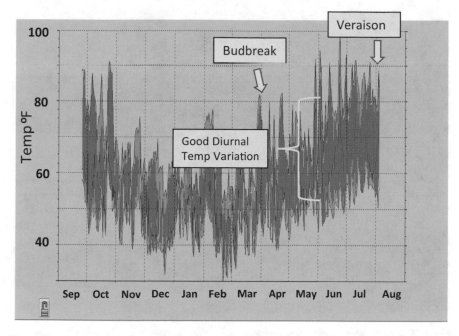

Fig. 6.27 Temperature highs and lows indicate reasonable growing heat conditions for Cabernet Sauvignon

The right ground water level is important to good viticulture practice. Water is key to vine growth and grape ripening. However, we do not want too much or too little. There is an exceptional compilation of valuable information on soil management, irrigation management, and nutrition management compiled and edited by Phil Nicholas (Nicolas, 2004). As discussed earlier, deficit irrigation is one important watering approach that can significantly impact fruit quality.

Understand your vineyard water drivers. Measure soil moisture as shown in Fig. 6.28. Decide whether to dry farm using mother nature's rainfall and or augment with irrigation. Incorporate rain, evaporation, soil type, and drainage to determine irrigation schedule. Compare with other local vineyards that produce quality grapes. Water for plant growth but do not overwater during peak grape growing months.

The sun and temperature environment are also a high likely suspect. For our canyon scenario, we are concerned whether temperatures will get warm enough to metabolize berries and provide a sufficiently long growing season for Cabernet. Will there be enough hot days to ripen our fruit. From our review of the temperature history, budbreak, and veraison, we can see that it does, but with little margin. We can also see that we have moderately good diurnal temperature variations.

We have discussed degree days earlier in the book. Particularly in reference to the Winkler scale. Figure 6.29 plots our scenarios accumulated degree days and shows that it compares favorably with a well-known growing region for Cabernet

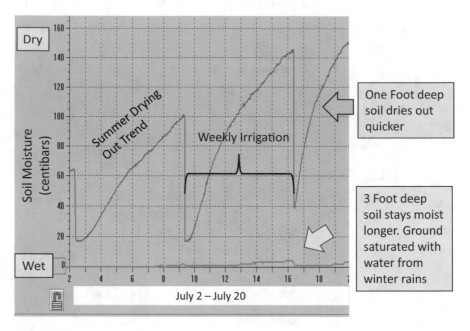

Fig. 6.28 Smart deficit irrigation: understand water levels at root depths throughout the growing season

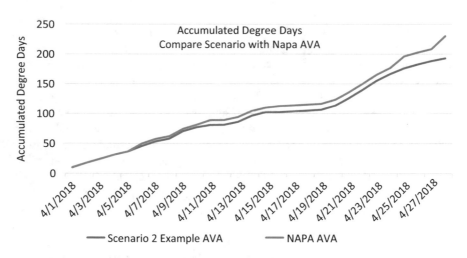

Fig. 6.29 Compare accumulated degree days with known good example, Napa AVA

Sauvignon. At least it did for this scenario's vintage. This is a vote to put it relatively lower on the list of possible imbalance causes.

Table 6.42 itemizes various preventive measures to mitigate our problems. We may not really ever eliminate our problems because our efforts will only be partially successful. Together, however, the combined effects will correct our problem for the next vintage. Managing our canopy is one such powerful preventive measure.

Table 6.42 Familiar preventive measures: trellis design, deficit irrigation, and canopy management

Possible causes contributing to imbalance	Preventive measures to reduce causes and improve balance
Viticulture: Grape growing conditions and/or varietal mismatch	• Deficit irrigate to reduce vine growth but minimal to ensure fruit ripens • Trellis orientation to maximize morning and afternoon sun • Manage canopy to minimize shoot growth and shade to grapes
Environment: Sun and temperature in the vineyard	• Trellis orientation to maximize morning and afternoon sun • Recommend VSP trellis design • Manage canopy throughout season. Wire train during early growth and prune during peak shoot growth before veraison
Material: High concentration of methoxypyrazine (MP) in Cabernet Sauvignon	• Same as above, trellis design and canopy management

Canopy management is necessary and critical to optimize growing conditions in any vineyard. Bad canopy management allows too many or too long shoots. This may shade other leaves and grapes from the critical sun. Too many shoots and or leaves will be a sink of vital nutrients, otherwise destined for our grapes.

Careful canopy management provides for 15–20 leaves per shoot and about 2 shoots per spur every 4–6 inches along the cordon. The canopy allows filtered sun on the grapes. These are approximate averages and may vary from site to site. Tracking cluster growth and development throughout the season and year to year provides good insight into what is best for your site and particular practice.

Figure 6.30 is a plot of temperature of grapes versus time of day illustrates the impact of canopy shading. This significant temperature difference or gradient is the kind of result discussed by A.A. Miller in the paper on the thermal regime of grapevines. (Miller, 1972). This effect can provide more hot hours in a day and more hot days in a year.

Let us now revisit the relative ranking of our three high likelihood causes of our balance problem as shown in Table 6.43. Which one of these is the most likely to be the root cause of our imbalance of vegetal to fruit characteristics of our wine? Given our current scenario with Cabernet Sauvignon grapes and their known high concentration of methoxypyrazine, we need to place this at the top of the list. The very nature of the grape material drives this kind of characteristic. This is not the case with other grape varieties. This essentially dictates the conditions to grow great grapes and make great wine.

The temperatures in the vineyard are comparable to known good Cabernet Sauvignon wine growing regions. This places the environment as the least likely of our highly likely suspects for the cause of our problem. Still a big or high concern, but at the bottom of our short "High" concern list.

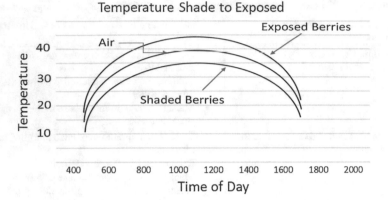

Fig. 6.30 Note canopy shaded grapes can be 10° cooler than exposed grapes

Table 6.43 Updated ranking of highly likely causes of our imbalance problem

Possible imbalance causes	Likelihood	Comments
Material: High concentration of methoxypyrazine (MP)	High plus	The Cabernet Sauvignon grape material itself has high vegetal concentrations. Dictates great care and high risk for greatness
Viticulture	High	Trellis design good with VSP and deficit irrigation applied, however, canopy management must improve for this crop in this vineyard
Environment: Vineyard sun and temperature	High minus	Vineyard temperatures sufficiently hot in comparison to known good Cabernet growing regions

6.10   RCA Summary of Red Wine Imbalance

Balance in a finished wine begins in the vineyard. Finding the root cause of an imbalanced wine also starts in the vineyard. Balance involves the relationship between wine components. The grape berry components will develop during the growing season. They will vary in relative concentration driven by the variety, terroir, vineyard management, harvest decisions, and winemaking practice. Table 6.43 listed and rank ordered our high likelihood suspects for our problem of imbalanced wine. We have not actually isolated a one true root cause; however, we have identified top driving suspects and for those we can influence, we should go after fixing them to improve the quality of next year's vintage.

We have presented a case of imbalanced fruit to vegetal characteristics in a Cabernet Sauvignon. This variety has its established natural components. An imbalance can occur with only slight variations. A thorough investigation of the vineyard and winemaking process, materials, equipment, people, measurements, and environment was presented. It is this careful hard work that is important to sort out the contributing factors and root cause.

Winemaking best practice with preventive measures is the path to making balanced quality wine.

References

Amerine, M. A., & Winkler, A. J. (1944). Composition and quality of musts and wines of California grapes. *Hilgardia, 15,* 493.

Boulton, R. B., Singleton, V. L., Bisson, L. F., & Kunkee, R. E. (2010). *Principles and practices of winemaking.* Davis, CA: Springer.

Cooke, M. G., & Berg, H. W. (1984). A re-examination of varietal table wine processing practices in California. II. Clarification, stabilization, aging and bottling. *American Journal of Enology and Viticulture, 35,* 137.

Crippen, D. D., Jr., & Morrison, J. C. (1986). The effects of sun exposure on the compositional development of cabernet sauvignon berries. *American Journal of Enology and Viticulture, 37,* 235.

Habertson, J. F., Hodgens, R. E., Thurston, L. N., Schaffer, L., Reid, M., Landon, J., et al. (2008). Variability of tannin concentration in red wine. *American Journal of Enology and Viticulture, 59,* 210.

Iland, P., Bruer, N., Edwards, G., Weeks, S., & Wilkes, E. (2004). *Chemical analysis of grapes and wine: Techniques and concepts.* Adelaide, SA: Patrick Iland Wine Production.

Mathews, M. A., & Anderson, M. (1988). Fruit ripening in Vitis Vinifera L. responses to seasonal water deficits. *American Journal of Enology and Viticulture, 39*(4), 313.

Miller, A. A. (1972). Thermal regime of grapevines. *American Journal of Enology and Viticulture, 23,* 173.

Mullins, M. G., Bouquet, A., & Williams, L. E. (1992). *Biology of the grapevine.* Davis, CA: Cambridge University Press.

Nicolas, P. (Ed.). (2004). *Grape production series number 2: soil, irrigation, and nutrition.* Adelaide, SA: South Australia Research and Development Institute, Winetitles.

Noble, A. C., Arnold, R. A., Buechsenstein, J., Leach, E. J., Schmidt, J. O., & Stern, P. M. (1987). Modification of a standardized system of wine aroma terminology. *American Journal of Enology and Viticulture, 38,* 143.

Peynaud, E., & Blouin, J. (1983). *The taste of wine: The art and science of wine appreciation.* (M. Schuster, Trans.). Paris: Wiley.

Smart, R., & Robinson, M. (1991). *Sunlight into wine.* Adelaide, SA: Winetitles.

Vivas, N., & Glories, Y. (1996). Role of oak wood ellagitannins in the oxidation process of red wines during aging. *American Journal of Enology and Viticulture, 47,* 103.

Winter, E., Whiting, J., & Rousseau, J. (2004). *Winegrape berry sensory assessment in Australia.* Adelaide, SA: Winetitles.

Chapter 7
Red Wine Color Resolution with Rational Decision-Making Methods

There is a well-known proven method for rational decision-making that can be applied to winemaking problem solving. This is the Kepner–Tregoe analysis (KTA) method. It was developed in the 1950s by Dr. Charles Kepner and Dr. Benjamin Tregoe (Kepner & Tregoe, 2013). It is a logical approach on how to think about a problem and its solutions. We will use the KTA method to evaluate and decide how to resolve our wine color problem.

7.1 Kepner–Tregoe Application to Wine Color Problem

Let us apply basic decision-making KT analysis to our wine color problem scenario. This method is a powerful tool to optimize decision-making incorporating real-world constraints. We will implement this in a way that uses your own personal real-world constraints. We intend to make a beautiful looking and tasting wine and at the same time, use this rational and practical approach to making the best decision about your winemaking process. As you know, every wine problem has a unique character-istic which may have been triggered at almost any time. Color of wine will vary with variety, blend, and age. Fig. 7.1 asks the question as to whether the color of your wine is what you want or what you expected. Discovering your wine color problem may be observed before, during, or after fermentation which will drive limitations on recovery options. You may be able to salvage this vintage by blending, but you need to be able to correct next year's vintage using a better approach. Different options will have different innate abilities to fix the problem. Prior methods have discussed finding the root cause and establishing preventive measures. We have multiple options on the table to correct the wine color. These options have different levels of impact on improving color and different levels of uncertainty about how well they will work. This scenario also incorporates the constraints of cost and schedule into the process. The severity will be established with respect to your own situation.

© The Editor(s) (if applicable) and The Author(s), under exclusive license to
Springer Nature Switzerland AG 2020
J. Steakley, B. Steakley, *A Quest for Quality Wine, Every Time*,
https://doi.org/10.1007/978-3-030-34000-1_7

Fig. 7.1 Is the color of
your Cabernet more like a
Grenache or does your
Pinot Noir look like a
Rose?

7.1.1   Quick Review of KT Analysis Basics

This is a rational process for deciding which winemaking process is best suited for
our wine style goal and our limitations on cost, schedule, and tolerance for risk.
Refer to Chap. 3 for a discussion of the basic KT analysis approach. Do not be
afraid to apply this rational and seemingly "dry and sterile" tool to a process that
will ultimately create a wonderful wine for us to enjoy and share with friends. The
two are not incompatible. This method and associated research will also take work,
but will ultimately save time, energy, and frustration by improving your chances for
success with the color of this or next year's vintage. This scenario provides us with
particulars of a given situation, but also illustrates how to think about your problem.
This is appropriate for any complex decision you face. The color of the wine in
Fig. 7.2 is not appropriate to the variety and style of our wine goals. In this tilted
glass, the color at the thin arc at the top appears to have a little too much brown hue.
As described in prior chapters, document your problem and unwanted color effects.
Include descriptive details of your wine's condition, your winemaking goals, and
wine color desires. Investigate and list recovery options, assign weights, and add up
scores. Compare the positives and negatives and choose the best-balanced path.
This may sound idealistic and simplified, however, these are the essentials and the
road we will take to learn this process.

Some of the earlier RCA method discussions in Chap. 3 involving cause and
effect diagrams and the five Whys can help with your basic research approach, asking the critical questions, and the documentation of your problem. We will repeat it
again: an important early step is to carefully assess your problem and situation. This
scenario has a problem with color. We are going to keep it simple and call it "too
light" for clarity of describing the method. In general, one should more thoroughly
describe your color and shades. Compare them with other wines of similar style.
Visual color is heavily influenced by the surrounding lighting conditions, the angle
of viewing, and the background behind the wine. You can tilt the glass against a
white background and see the shades of red and how they also vary with the thickness of the wine. You may also want to get others to give their opinions or if you
have the resources, make spectral measurements to repeatably and accurately quantify your condition. However, for this scenario and our desire to illustrate the
decision-making aspect of this method, we are going to simply state our problem as

Fig. 7.2 Does the color of your wine have shades that just do not look right for your style?

we want a darker red color. This KTA step is often applied after solution options are identified and you are ready to make your decision.

This step in this example will construct an integrated matrix using positive and negative characteristics in essentially one step. The more expanded approach calls for a positive characteristic assessment and its own matrix followed by an effort to create an adverse or negative characteristic assessment with its own separate adversity matrix. Both approaches are the same. They add up positive characteristics and subtract negative adverse effects to give the most balanced decision. In most business decisions, it is important to consider cost, schedule, and risk. Cost and schedule aspects might also be considered constraints or limitations given your budget and schedule. If you have a lot of money, the budget constraint may be low. In this scenario we have some money and resources to apply so the cost issue is moderate. If you discovered your problem early in the production process, the schedule constraint may be low.

We discovered the color issue after fermentation and pressing, and partially into the aging process. We will not be able to fix this vintage except by blending, but we should apply corrections next year. However, we do have schedule pressures of getting our wine to market, so we still consider the schedule required to correct our process of moderate concern.

Risk is the uncertainty that goes along with all real-world actions and consequences. We really want the red color to improve and have moderately pressing cost and schedule concerns so when we establish relative weights, this fourth risk category will not be zero, but it will be smaller than the others.

We will select one of three options, while considering their respective impact on color quality, cost, schedule, and risk. The intent is to make the most balanced and optimal decision.

7.1.2   Define the Color Problem Resolution Options

Let us review the problem and frame our options to fix it. Our problem is wine color. Improving the quality in this scenario is to make the color a darker red. Our decision will be to pick the best option. We know wine color, whether it is red or white, may have a shade or tint or level of clarity that does not really fit the style of wine we expect. Our current example is based upon a quality problem with the color of our wine. We wanted a dark red Cabernet and ending up with a light red Cabernet. We will study the problem and identify possible solutions or options to fix it now or fix it in next year's vintage. Table 7.1 describes features of three possible resolution options. Each of these options will have issues and risks associated with implementing them. We want more color in our wine, so one of our most likely cause is lack of time on the skins. Given that the root cause is obvious, we do not need to apply the rigorous cause and effect analysis or other methods; so here we will illustrate KT Analysis for decision-making of the best solution to this problem. Increasing time of the juice on the grape skins will increase the color. Let us discuss the options to increase time on the skins.

We have a problem with the color of our wine, and we have three possible options to fix it. How do we choose which option is the best? We need to think and define what we actually mean by the best decision. At the same time, we need to think

Table 7.1 After color quality problem observations, research, and consultation; we have identified three options

Option 1: Add cold maceration before fermentation for more skin contact time	Option 2: Increase time in fermentation through yeast selection	Option 3: Increase fermentation time with lower temperature
(a) Known process which directly increases contact time with the skins (b) This can be implemented by keeping the juice on the skins cold and longer before fermentation. (c) Ability to track time by following days and hours is very accurate (d) Risk to excessive air exposure (e) Minimize keeping crushed grape skins and juice cold	(a) Known process which can increase skin contact to impart more color (b) More time is realized by lengthening fermentation with less aggressive yeast (c) Added benefit of higher temperature to improve color extraction (d) Risk because dominant yeast type, conditions, and rates are uncertain (e) Moderate time uncertainty due to unknown process duration	(a) Known process that will increase skin contact time (b) More time is achieved by lowering the fermentation temperature and thus the rate. (c) Potential unintended consequences from lower temperature (d) Moderate uncertainty in process duration

about how our options might affect these kinds of characteristics. This can be efficiently done in an iterative way. Breakout and clarify each option, then breakout and clarify your constraints and limitations, then go back and update details of your option in ways that relate to your desired attributes. How well does each option actually affect color intensity and how well can each option be controlled? We care about cost and schedule, so figure out how much each option costs and how long does each option take. Are there risks and uncertainty of each option in achieving our improved color goal or risks resulting in unintended consequences? Do we have a large volume of wine that is in jeopardy and a big risk to losing lots of money in sales, not to mention a hit to credibility by producing low-quality wine? We will be evaluating these and assigning relative numerical scores later so it is helpful to understand the details of these aspects. By understanding the details of our options, we understand how they might impact our attributes and make a better decision.

You now have defined and documented the details of your various options. Our next step will be to define attributes of our decision that are important to us. For this scenario we have chosen quality, schedule, cost, and risk. This scenario's quality issue is the lack of wine color. This example scenario has the one quality problem, however, note that this method can be extended to include any number of quality problems simultaneously. The categories are intended to be specified and applicable to your own problems, conditions, and goals.

7.1.3   Define Color Decision Attributes and Assign Numerical Weights

To begin the KTA process, we will create the attributes that we want to evaluate in our decision-making process. Table 7.2 shows us example attributes, weights, and background comments. These attributes or categories will typically include the quality characteristics we want to improve. This is the problem or quality shortfall. The overall decision will also include the cost and schedule impacts of taking action to fix and prevent the poor quality from occurring again. We will also include an element of risk. This is the risk that an unintended consequence may occur. As we all know, some level of risk is associated with the real-world decisions. We may not achieve the outcome we want.

Next, we will assign weights and scores to these categories in accordance with our values, situation, and options to resolve it. In this example, we are going to use percentages to assign relative weights. You may also use weights and then divide by the sum of all the attribute weights to get a fraction. This percentage approach helps us think about how the weights compare to one another and relative to 100% or something that represents all the considerations. One can get a quick sense of the relative portion or weights by noting rough fractions or percentages of each. Sometimes we only have a rough feel that half of this decision should be on fixing the problem and the other half on cost and schedule. Percentages give you a quick-look assessment of

Table 7.2 Pick attributes or characteristics of the problem solution that are important to you and assign them relative weights

Item	Attribute	Weight	Comment
1	Quality shortfall: Color	40%	• Darker color is the quality attribute to improve • Our Cabernet Sauvignon is too light and looks more like a young Grenache or Pinot Noir • Color is the first or second thing one notices about a glass of wine • Assign a numerical weight for its relative importance
2	Schedule	30%	• Schedule is commonly used in business decisions • Fixes or preventive measures will take time • The weight will depend upon your values and situation • If discovered well ahead of wine production plan, then less important, and assign relatively lower weight • In this case we do not have much time to improve color and may have to wait until next vintage
3	Cost	20%	• Cost also used in business decisions • The cost of different solutions will vary • If tight budget, then cost can be very important and should be assigned relatively high weight
4	Risk	10%	• Risk reflects the uncertainty and complexity of solution or preventive measure • Wine production is very complex, and all actions may have unintended consequences • If we tolerate high risk processes in experimental wines, then this weight should be low to not impact decision. If production wines, then relatively higher
		100%	• Weights must add up to 100% to represent complete decision process and all factors important to you

this. Let us next breakdown the color quality issue and review ways to think about this category.

Quality is our goal and our wine color falls short of our expectations and reasonable standards. As we covered in the quality section, there are many different quality perspectives and proxies for what quality means to you. Think about your own definition of quality and what your goal is to improving it. This quality or performance improvement is what we want to include in this line item. You may have multiple quality characteristics that you want to improve in this particular decision or impending action ahead of you. When you are delineating these, too many separate quality line items will make it confusing to organize and blur the ability to address each. However, if there are a few top key attributes that you want to make sure you include in this decision, then include them. We have a top-level sense that we want half our decision on achieving a darker color and half on schedule and cost. We are going to allocate 50% to the color attribute and associated risk. We want the positive color attribute to get the majority of this share so we will break it down further by assigning 40% to the color quality and 10% to associated risk.

Many wine characteristics are coupled and likely to be connected when a particular corrective measure is executed. It is difficult, and may be close to impossible, to apply a change and impact only one thing and not affect others in some way. You may want more color, but not want to severely impact the mouthfeel of the wine or some other feature. If you are worried about this and have strong preferences on certain qualities, then include them and do the appropriate work to understand them and include them in your decision-making process. In this first example, we have just one category and that is the quality feature of color. Our problem is our red wine is too light and we want a deeper red. So, as we compare each option, we will score their effectiveness at producing a darker color.

The next category is schedule. How long does it take to implement the option? This reflects the duration or time it takes to change the process to incorporate the new option. Short duration preventive measures with the same effect are better than long duration preventive measures. We hope and want easy quick fixes, however, brace yourself; these are rare. Short duration fixes are rarely the case in viticulture and vinification. Think about your particular quality problem, when it was first discovered, and when in the production cycle it is occurring. Also think about the recovery or preventive measure options you are considering. Is the fix schedule more or less than the time you have available? As part of assigning a relative numerical weight, also consider its relative importance with respect to the other categories. We already have assigned 50% to color quality and risk. That leaves schedule and cost. We think schedule may be a little more important to our decision than cost, so we will tentatively assign it 30% leaving 20% for cost. These are preliminary assignments and we will come back and review them after we get through all the categories and assignments.

Cost is the next category. Figure 7.3 presents an example of formulating a weight associated with the price of oak barrels. In the real world we live in, there is always a cost to implementing a change. This cost can be either the direct financial expenditure like the cost of oak barrels or the labor cost of employees like winery staff or vineyard crew. For other scenarios, there may be a wide range of costs. Some of these might be the cost of new grape vines, watering, fertilizers, sprays, commercial yeasts, or electricity for air conditioning the winery. There are many different labor and material costs. Some of these costs may be one-time non-recurring charges or some may be recurring. If you need to plant new grape vines, the cost of the grapes could be considered a one-time charge. However, there will be also be annual recurring costs to maintain, water, fertilize, and manage the larger canopy. Take some time to think through all these costs to get a good estimate.

We already have a preliminary weight for this cost category of 20% given that we established all the others. Part of our iterative review process is to confirm this is reasonable or adjust it. After you have got a good cost estimate for each option, compare with your available budget. If you have the budget to cover these options, then no need to update the relative weight. If it is close but within the uncertainty of your estimates, this might mean this cost attribute is more important and you might want to adjust the numerical assignment up and drop one or more of the other categories to keep the sum at 100%. Also, if an option cost is way above your available budgets, then it does not meet your hard requirements and should be dropped from the list of possible of options.

Fig. 7.3 Weight your attributes in accordance with your near- and long-term plans. You may have recently bought new barrels and do not have the budget for new equipment this year, so cost is assigned a relatively heavy weight

The fourth and last attribute in this example is risk. At this point, you will have researched various options for improving your poor-quality color. Each of these will have uncertainties and will only be partially successful. This uncertainty in completely fixing or preventing the problem is represented by this risk category. If one of your options directly adds more time on the skins before fermentation but you do not use sulfur, then the risk is on the high side due to potential spoilage. If one of your options is more in the experimental stage and uses various new enzymes or other untested chemicals and conditions, the risk should be higher. Including the category is a good consideration in accounting for uncertainty. We have a preliminary assignment of 10% weight for this risk category. Review your level of understanding of the problem and the context of your wine production situation. Our scenario is considering fairly well understood processes so we are not going to adjust our 10% numerical weight score.

We have made preliminary weight percentage assignments to our four attributes. The sum of the weights must add up to 100%. This makes good sense because 100% represents everything you value in making this decision. If you only cared about two things, color and taste, weight each according to importance so they total 100%. If you had no cost and no schedule concerns, this might work. However, if you want this decision model to reflect your budget and time available, it is important to include a cost category and schedule category. The complexities of our world make it advisable to include a risk category to represent some uncertainty in whether chosen option will work or not.

It is your own judgement on how to assign the relative magnitude of your categories. After you have taken your first cut, review it again. If these four attributes were each equally important to you, or you do not have any particularly strong reason to weight one over the other, you might assign equal weightings. This would mean color quality is assigned 25%, schedule is assigned 25%, cost gets 25%, and risk is assigned 25%. They should add up to 100%. Each year, the relative weights of what is important or critical to your decision may also vary. Take the time to dwell on each attribute category and how it fits into your values, problem, and situation. This goes directly into your decision.

7.1.4 Assign Numerical Scores to Color Fix Options and Attributes

We need to assign a numerical score to each category for each option. It should be based upon our estimate of the likelihood of the option to improve or increase the respective attribute. Table 7.3 shows our three options and comments on their respective features and issues related to changing your wine color. Will the option work? How well will it work? Think about its chance of working. What is your confidence in this option improving the situation? Will you be able to implement the option and control the environment for it to work properly? If the option will surely work, then you may assign it a 9 or 10. If option may or may not work, it may only get a 5 or less. Think about the complexity of your solution options and your ability to control the variables. If a solution involves changing the chemistry, do you have a good understanding of the relative concentrations or good control of the time and temperature? If the option involves split second timing or very precise temperature

Table 7.3 After the options are identified, think about their respective likelihood of improving the color quality

Option 1: Add cold maceration time for more skin contact time	Option 2: Increase time in fermentation through yeast selection	Option 3: Increase fermentation time with lower temperature
(a) The problem is evident by visible color flaw (b) More time on red grape skins will impart more color (c) This option is a good way to get more skin contact time by additional cold maceration time before fermentation (d) Controlling the time and checking the color progression progress is reliable	(a) Option 2 is another way to get more time by lengthening fermentation by using a less aggressive yeast (b) Yeast type choices are available, but controlling conditions can be challenging with moderate uncertainty in results (c) Change in skin contact time is relatively small compared to option 1 so color change is also expected to be smaller, but temperatures may will be higher driving more color change	(a) Option 3 is another alternative to get more time by lowering the temperature and expending the rate to slow thus increasing skin contact time (b) Implement with temperature-controlled fermentation tanks (c) Still moderate uncertainty of results due to unknown changes in rates of reactions

control or many unknowns, it should get a low score. This step of assigning a numerical score is about the relative likelihood of a particular option to improve the color quality attribute, or the relative effectiveness of each.

Table 7.4 describes how to establish a numerical score for each option. If we estimate there is a very high likelihood of success and impact, then the assigned score should be close to the max of 10. If we do not have much objective data at all, the score could be much lower. We are also choosing between three different options so we want to separate our numerical range accordingly. We want the most likely to get the highest score, around the max of 10 and we want the least likely option to be assigned a low score close to 1. In this KTA method, the absolute value of this scoring is not critical, the relative value is.

We have shown scoring assignments for our color quality attribute in the above table. Note that they are numerical, but the magnitudes are only approximate, and they are only meaningful in their relative ranking with respect to the other options. By assigning the most likely close to 10 and the least likely close to 1, we are spreading out the integrated sums but not hurting the logical integrity of the process. Ultimately these scores are multiplied by the weights and the product or weighted score will be summed to be assess for the best choice.

Now we will assign numerical scores for the schedule attribute for each of the three options. Table 7.5 describes how this is done for this wine color problem and our three options. Think carefully about the time to implement each option. If you add more cold maceration weeks, that is a direct addition to time to produce the wine. If you add more time during fermentation by using a less aggressive yeast or lowering the temperature, this will also add time to the production process, however, it may be a smaller portion of time. These should be estimated and included. This will be important in comparing your options and making your decision. Note that you may have so much schedule pressure because you are already behind, that

Table 7.4 Assign numerical score from 1 to 10 in accordance with likelihood of achieving quality improvement

Option 1: Add cold maceration time for more skin contact time	Option 2: Increase time in fermentation through yeast selection	Option 3: Increase fermentation time with lower temperature
(a) Directly adding skin contact time brings high confidence in imparting color (b) Low residual uncertainty in amount of color to be imparted (c) We expect to add relatively long durations of skin contact time (d) On scale of 1–10 and compared with other options, we will assign it a 10	(a) Likely time increase, but uncertain outcome due to complex yeast reaction characteristics and wine conditions (b) Likely increase time and skin contact, but delta length is small compared to option 1 (c) On a scale of 1–10, our confidence is much lower than option 1, so we are going to assign a score = 2	(a) Confident that lower temperature will slow reaction time, however, yeast sugar conversation rate has uncertainty (b) Also, small time increase compared to option 1 (c) On a scale of 1–10, the result is about moderate confidence so we are going to assign a score = 4
Numerical score = 10	Numerical score = 2	Numerical score = 4

Table 7.5 Assign scores to schedule attribute of each option. Scores range from best or shortest (10) to worst or longest (1) for this group of options[a]

Option 1: Add cold maceration time for more skin contact time	Option 2: Increase time in fermentation through yeast selection	Option 3: Increase fermentation time with lower temperature
(a) Cold maceration time before fermentation can be typically many weeks (b) Typically longest period relative to other options. (c) Set to min score or worse for this group of options. Assigned score = 1	(a) Differences in rates of yeast sugar conversation to dryness typically vary a few days (b) Relatively small schedule change compared to option 1 (c) Comparable in skin contact time change in schedule duration as option 3 (d) Assigned score = 5	(a) Can control temperature and thus influence time, however, also typically change fermentation duration a few days (b) Similar schedule duration change as option 2 (c) Assigned score = 5
Schedule score = 1	Schedule score = 5	Schedule score = 5

[a]Note in this scenario using the KT approach, we are combining positive quality characteristics with adverse negative aspects into one KTA matrix, so the adverse scores will have 1 (most adverse) to 10 (least adverse) assignments

some fixes or preventive measure options may have to be put off until next year. Adding cold maceration time by implementing option 1 will add a range of many days to weeks on your production schedule. Increasing fermentation time by either option 2 or 3 will add a few days to your production schedule, but not weeks. So, we are going to roughly assign the same value of 5 to both option 2 and 3 compared to a 1 for option 1. Remember, more schedule is an adverse impact so it gets the lowest score for the way we are adding up scores in this example. In the end, the highest score wins so adverse things should get relative low scores.

Now we are going to assign scores to the cost attribute of each option. Table 7.6 describes cost influencing features and assigns a score for each option. For cost, an adverse attribute, we assign a low score for high costs and a high score for low costs. Option 1 has the expected highest cost largely because of its lengthy additional time to the recurring and sustaining production costs. Options 2 and 3 will impact days of production schedule which is considered small compared to the weeks of option 1. We consider them similar in cost for this scenario because of the assumed no significant materials or new equipment needs. We will assign them each the same value of a score of 5.

Let us discuss risk and how it relates to our options. Table 7.7 describes risk features and how to formulate a risk score for each. Recall that risk itself is a difficult thing to assess. It is the risk of uncertainty or the unknowns of what might happen. In order to bound these risks and assign a number, it requires some knowledge, experience, and or track record. Our options vary the time on the skins, however, are implemented in different ways. This brings different uncertainties and risks. Watching the wall clock and timing duration in cold maceration is less risky than adding an organism to the grape juice and understanding the effect. A commercial yeast is developed and will likely have predecessors, testing, and some track record.

Table 7.6 Assign cost scores to options. Note high cost is bad so it should get a relatively low score and low cost is good so it should get a relatively high score

Option 1: Add cold maceration time for more skin contact time	Option 2: Increase time in fermentation through yeast selection	Option 3: Increase fermentation time with lower temperature
(a) Cold maceration is relatively long cold period before fermentation so costs include recurring air conditioning utilities and spot check labor for sustained period (b) Long period recurring costs relatively large for this group of options (c) Assume no new non-recurring equipment nor winery costs for this example (d) Relatively large expense so gets relatively low score and is assigned score = 1	(a) Costs of alternate yeast probably not much change assuming commercial yeast (b) Costs of short fermentation period incurs recurring utilities and spot check winery labor relatively small compared to option 1 (c) Assumes no new non-recurring equipment or winery costs (d) Relatively low costs so this gets relatively high score and similar to option 3 (e) Assigned score = 5	(a) Costs of lower temperature implemented through existing equipment so no non-recurring new equipment costs (b) Costs for additional utilities and spot check labor small compared to option 1 but like option 2 (c) Assigned score = 5
Cost score = 1	Cost score = 5	Cost score = 5

Table 7.7 Assign risk scores relative to the three options. Note that the higher risk options get the lowest or worst score

Option 1: Add cold maceration time for more skin contact time	Option 2: Increase time in fermentation through yeast selection	Option 3: Increase fermentation time with lower temperature
(a) There is high certainty of imparting more color with more skin contact time the risk is considered low. (b) This is the lowest risk of these options because it results in the longest skin contact time. (c) Assign max score of 10	(a) Commercial yeast has fairly well-known transformation rates, however, your conditions, yeast available nitrogen (YAN), and your juice conditions can be uncertain (b) Natural yeasts may also have unknown compositions that add uncertainty to fermentation rates (c) This risk is higher than option 1 (d) Assign relative low score of 2	(a) Risks of increasing rate change through temperature change are moderate (b) Lower temperature means lower rate, but the exact scale factor is uncertain. (c) This option is considered less risky than option 2 but more than 1. (d) Assign moderately low score of 4
Risk score = 10	Risk score = 2	Risk score = 4

A natural yeast from the vineyard has many unknowns that vary with year to year weather and pests. At first blush, a commercial yeast is less risky than a natural yeast. However, if you are experienced with a vineyard and making wine with yeast from that vineyard, then this may also be relatively less risky. The message is to consider all the factors in your own situation.

The least risky option is to directly increase skin contact time by adding time in the process. The time aspect can be readily implemented. This option typically adds

many days if not weeks of skin contact. There are potentially unintended consequences due to air exposure, however, if kept cold, this can be very effective. We
consider this the least risky, so it gets the highest score of the three options. Option
2 involves changing the yeast type. The yeast type will bring its own requirements
on the fermentation process and the net delta increase in fermentation time may be
very uncertain. Relative to option 1, this is much riskier and therefore gets a relatively low score. Option 3 assumes a lower temperature will result in a lower fermentation rate. There are potential risks with stalling or a stuck fermentation, however,
for this scenario we will assume this is controlled and proceeds normally. We consider this a moderate risk and assign a low value between the other two options.

You should always consider your own problem, situation, knowledge, and experience. You should also use your team's or other outside expert advice whenever
needed. As we mentioned, typically commercial yeast is more understandable and
controllable than natural yeast. However, this is not always the case. You may have
a good track record with natural yeast from a known vineyard. You may also have a
good history of making quality wine with this natural yeast. Patrick Newton is a
very talented chief winemaker at a successful commercial winery in New Zealand.
He makes wonderful wines using natural yeasts and makes high-quality well-
balanced wines.

Now we have documented our problem, listed attributes, and defined options to
correct our problem. Now let us use our scores, calculated weighed values, and create our integrated KTA matrix.

7.1.5   The Integrated KTA Matrix and Best Decision

Let us now assemble and build our integrated KTA matrix. We have evaluated our
situation, assigned scores, assembled the building blocks of our decision-making
KTA matrix. The major columns are the options and the rows are the attributes.
Each attribute has an associated fractional weight, each option has a raw score for
each attribute, and the weighted score is the product of the two. Assemble this for
all the options and for each attribute. The sum of the weighted scores for each attribute is the integrated weighted score for the option. The highest integrated weighted
score is the best choice. This is what we consider to be the most rational choice. It
is the most logical and sensible decision.

This integrated matrix, shown in Fig. 7.4, is a concise visual way to collect and
display your characteristics, options, scores, and relative rankings to see the best-
balanced decision. In the more formal extended version of KT analysis that we
described in the basic method chapter, we created two separate matrices, however,
both methods are functionally the same. The more extended approach created a
positive quality matrix and a separate adversity matrix and subtracted the two.
However, in this scenario, we used one simple positive characteristic of color and
the adverse business attributes of schedule, cost, and risk. Any adverse attributes
were essentially assigned a reverse score. This means that the most adverse got a

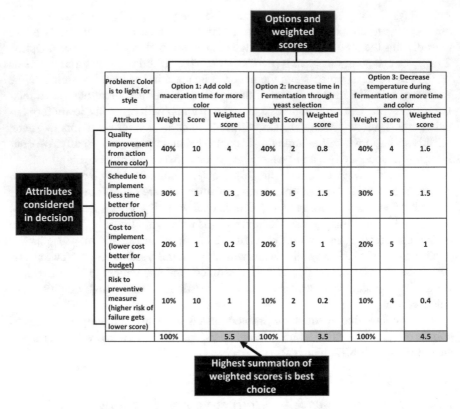

Fig. 7.4 The Integrated KTA matrix forms weighted scores for each option. The option with the highest sum of weighted scores is the best choice. Option 1 is the best choice and option 3 is the worst

low score and the least adverse got a high score. The longest schedule is the most business adverse and is assigned a one and the highest cost is the most adverse and is assigned a one. The riskiest option was assigned the lowest score and the least risky was assigned a relatively high score. Note in the graphic integrated matrix for this scenario shows option 1 gets the highest score and is the best choice.

7.1.6   Sensitivity to the Magnitude of Your Weights

Let us take a look at how the magnitude of weights affects the decision. It is very clear that when one has one variable and nothing else matters, then the decision is based upon maximizing that one variable. If one wants to improve the color quality and you have all the time and money you need, one can decide on maximizing color improvement. However, as we have discussed, decisions are made in the context of what is important to you as a winemaker, or maybe in accordance with the owner's resources, or maybe with what markets dictate. Different values can be incorporated into the process. This KTA approach allows these separate attributes to be uniquely weighted. This essentially ranks them in order of importance. This feature becomes a part of the decision-making. We recognize that every problem and solution will vary in detail for every unique situation. We now have a tool that also recognizes and accounts for your own changing values and circumstance. As a winemaker or wine producer grows or changes, aspects of this growth can be incorporated into new weights of what is currently important. These can then be applied to your current quality problem and solution to ensure you are making the right decision.

We will use this color quality example to illustrate the sensitivity to the weights of our attributes and show the results in Fig. 7.5. We can use values that are heavily weighted in favor of the quality parameter, or we can emphasize the schedule weight, or we can emphasize the cost weight. These are features we can adjust and tailor to our situation. There are three different weight profiles and decision results for the same problem and situation. Only the decision attribute weights have changed. Note how the decision changes.

For the same quality color problem and situation, the optimized decision changes because the values changed. For the case where color quality was of the highest value, option #1 was selected. For the case where schedule was of the most value, option #3 was chosen. For the case where cost was most important and the same problem situation, option #2 was chosen. These are just example values used to illustrate the effects. In general, you decide what is important. You apply your own values, weights, and scores to arrive at the best decision.

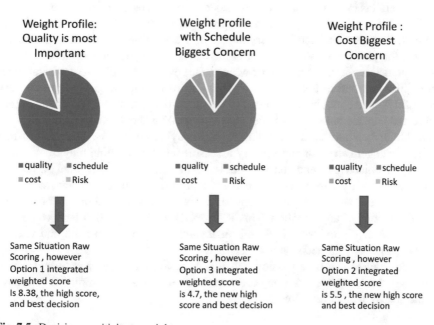

Fig. 7.5 Decision sensitivity to weights

Note the detail breakout in the following three corresponding tables. Note how the weighted scores stack up. Fig. 7.6 shows the detail breakout for the case in which quality is the most important and option 1 is the best decision. Fig. 7.7 shows the detail breakout for the case in which schedule is the most important and option 2 is the best choice. Note in Fig. 7.8 that cost is the most important and option 2 is still the best choice, but option 1 is now a distant third choice.

Remember all of these are for the same color problem and the same three options to repair or prevent it from happening again. Again, each are for different profile weights. One is heavily weighted on quality, the second is heavily weighted on schedule, and the third is on cost. The situation raw assessment is the same for a given option, but the best-balanced decision and worst decision is different for different weights. This of course, is as it should be. For a given problem and when you have different quality desires and constraints, the answer could very well be different.

7.1.7   Summary of the KT Method Applied to Our Wine Color Problem

We made use of the KT analysis approach in this wine color quality problem to determine the best option to fix it. This is a rational method that organizes and optimizes your own values in making the best decision. It can be tailored to any number of quality parameters to incorporate in the decision. It will also take into account adverse effects or unwanted consequences to make it a balanced decision.

This KT approach helps clarify many contributing, potential confusing, and complicating factors faced in every problem. Discovering a problem at the beginning of the growing season or just after harvest can give you more options and time to fix something in the vineyard or winery. Discovering a problem at critical junctures like just before bottling and distributing to paying customers can be very stressful. You may not have the time nor resources to fix it. It could mean the regrettable and considerable loss of a valuable vintage. You should make sure your wine is stable before bottling and sending it off to paying customers. That cost of that extra wine chemistry test and sensory assessment before bottling may be worth it. Having an objective science-based assessment before bottling may head off distributing poor quality wines. It may head off a big hit to your reputation for quality wines. This KT approach can account for the many unique conditions, constraints, and limitations of any situation. This KTA method not only helps lead you to the best decision, it provides a tool and mechanism for you to go back and review, explain, and vet your decision to others.

Making wine can be a passion and joy, however, the journey may look bleak when you hit big problems along the way. Use this KT approach on your quest to make quality wine, every time.

Problem: Color is to light for style: Quality heavy weight	Option 1: Add cold maceration time for more color			Option 2: Increase time in Fermentation through yeast selection			Option 3: Decrease temperature during fermentation or more time and color		
Attributes	Weight	Score	Weighted score	Weight	Score	Weighted score	Weight	Score	Weighted score
Quality improvement from action (more color)	80%	10	8	80%	2	1.6	80%	4	3.2
Schedule to implement (less time)	14%	1	0.14	14%	5	0.7	14%	5	0.7
Cost to implement (lower cost)	4%	1	0.04	4%	6	0.24	4%	4	0.16
Risk to preventive measure (higher risk-lower score)	2%	10	0.2	2%	4	0.08	2%	2	0.04
	100%		8.38	100%		2.62	100%		4.1

Fig. 7.6 This KTA matrix has the highest quality weight. Note how weighted scores stack up for given raw scores in favor of option 1

Problem: Color is to light for style: Schedule heavy weight	Option 1: Add cold maceration time for more color			Option 2: Increase time in Fermentation through yeast selection			Option 3: Decrease temperature during fermentation or more time and color		
Attributes	Weight	Score	Weighted score	Weight	Score	Weighted score	Weight	Score	Weighted score
Quality improvement from action (more color)	10%	10	1	10%	2	0.2	10%	4	0.4
Schedule to implement (less time)	75%	1	0.75	75%	5	3.75	75%	5	3.75
Cost to implement (lower cost)	5%	1	0.05	5%	6	0.3	5%	4	0.2
Risk to preventive measure (higher risk - lower score)	10%	10	1	10%	4	0.4	10%	2	0.2
	100%		2.8	100%		4.65	100%		4.55

Fig. 7.7 This KTA matrix has the highest schedule weight. Note how weighted scores stack up for given raw scores in favor of option 2

Problem: Color is to light for style: Cost heavy weight	Option 1: Add cold maceration time for more color			Option 2: Increase time in Fermentation through yeast selection			Option 3: Decrease temperature during fermentation or more time and color		
Attributes	Weight	Score	Weighted score	Weight	Score	Weighted score	Weight	Score	Weighted score
Quality improvement from action (more color)	10%	10	1	10%	2	0.2	10%	4	0.4
Schedule to implement (less time)	5%	1	0.05	5%	5	0.25	5%	5	0.25
Cost to implement (lower cost)	80%	1	0.8	80%	6	4.8	80%	4	3.2
Risk to preventive measure (higher risk-lower score)	5%	10	0.5	5%	4	0.2	5%	2	0.1
	100%		2.35	100%		5.45	100%		3.95

Fig. 7.8 This KTA matrix has the highest cost weight. Note how in favor of option 2, but option 1 is now a distant third for same raw assessment

7.2 "Quick-Look" Rational Assessment

Not all decisions require a protracted full-blown KT analysis. Some can be solved with a quick rational assessment of your situation and options. Sometimes you do not need to get at the actual root cause, you just need a timely good decision. Sometimes you can work out the root cause later. This is a "quick-look" or rapid rational approach for just such a situation. This wine color problem was discovered during aging and prior to bottling. We still have a color problem, but it was discovered late just before bottling and the truck is coming soon, so there is a sense of urgency to make a decision. The timing is different than the previous situation. The above scenario was to be addressed in planning for fermentation and cold maceration. This time, you realize, during the aging phase, that the color is not what you want. You might be getting relatively late direction from your boss, or sales, or your customers. They may have expressed an interest that they prefer or expect a darker red or more purple shade of wine. Let us look at a quick rational approach to resolve this. Table 7.8 is a set of questions to ask and answer. Answer what you know, but still list what you do not know or what information is missing.

Assess the situation and layout what you know and what you do not know. Answer the general questions about what you know, when was the problem observed, where was it observed and how bad is it. Collect the facts and review them. You know the color is not right because you can see it. You have seen or compared with other quality wines and recognize the color problem. Take a few samples from different barrels, ask your assistant or trusted advisor for a second or third

Table 7.8 Quick-look questions. Answer with what you know and what you do not know

Question	What you know	What you do not know
• What is the problem?	• Color too light • It is a light red, bright, and almost clear • It is not as dark as you like, nor what your clients expect • Aroma and taste are ok	• Do not know why
• When first observed?	• During aging • After about 17 months, 1 month before bottling	• No color info at start of aging • Not clear if missed color inspection after fermentation
• Where did it occur?	• In the barrels from Hillside vineyards • Grapes from block #12	
• How much	• Ten barrels impacted	
• To what extent?	• Color better than last vintage, but not dark enough for desired style • Not the darker shade of red and purple hue you wanted • This problem is not in all 100 barrels for this vintage. • Limited to ten specific barrels from vineyard block #12	

opinion. Describe and document how the color is not what you want. You can compare a photograph with a prior vintage or standard color palate.

The color is not right, but is the aroma and taste still good? Do these other characteristics meet your quality goals? Is the color the primary feature you want to adjust? As you are thinking about adjustments, just before bottling, you have limited options. You also do not want to adjust the color and at the same time, worsen the taste or aroma. Inspect, smell, and taste samples. Make additional independent assessments. Many people may not have the experience or training to evaluate the quality of a wine by its aroma or taste. However, many will have a better chance to distinguish between two different wine blends when evaluated side by side as shown in Fig. 7.9. One can focus on comparing one aspect at a time between two different blends. Is the wine in the left-hand glass redder than the wine in the right-hand glass? Is the aroma from the left-hand wine fruitier than the right-hand wine? How would you describe the taste difference between the left-hand wine and the right-hand wine? Which do you like better? Why? Include multiple evaluation techniques for better feedback on your blends.

Evaluate the chemistry. Measure the wine pH and acid and make sure it is within stable ranges. Ask yourself when the color problem was first observed. Note when in relation to the various phases of winemaking. In this case, it was first observed during aging. Breakout the timing more precisely. Which month of the aging process was it detected? Was this aging planned for 12, 18, or 24 months long? Color inspection notes taken after fermentation and before aging were ambiguous and suspect. Take a note to self to make process more robust next vintage. What is the planned schedule for bottling this year? How many more months until bottling? Do

Fig. 7.9 Even when blending to improve color, include evaluations of aroma and taste. These assessments can be more reliable with multiple people assessing wines side by side in two separate glasses

you have any flexibility in the bottling date? Can it be rescheduled? Clearly establish how much time you have got before bottling. How much time do you really have to fix this color problem?

The "where" and "how much" part of this assessment continues with determining which barrels and/or tanks have this color problem. Sample to understand which barrels and/or tanks are impacted. Different barrels may hold different wines from different vineyards or different vineyard blocks. Each of these may have their own color characteristics. Not all may need correction. Inventory and identify which barrels or tanks are impacted. Determine what volume of wine needs adjustment.

Blending is an option you may consider. Fig. 7.10 shows a possible option for blending with a darker variety. Blending a darker wine with a lighter wine will result in a blend darker than the original. Seems obvious, but do not forget there will be taste changes too. It is essentially becoming a different wine. The relative concentration of the two will determine the final integrated blend color. Adding a little inky dark Petite Sirah to a lighter Zinfandel, Syrah or Cabernet will darken the color of the original varieties. The blended Petite Sirah is very dark so a little goes a long way. There is a point at which the flavor and aroma of the Petite Sirah will interfere

Fig. 7.10 Dark "Inky" Petite Sirah is candidate for a blending wine that darkens color. A small percentage may be all that is needed to darken color enough

Table 7.9 Increase percentage of Petite Sirah to darken color of blend

Blending trial samples		
Blend #	Zinfandel	Petite Sirah
1	100%	0%
2	90%	10%
3	75%	25%
4	0%	100%

or start to dominate the flavor and aroma of the blend. We are expecting the color will darken noticeably before the aroma and taste are impacted.

This situation calls for blending trials. Table 7.9 lists various possible combinations of different percentages of Zinfandel and Petite Sirah for our blending trial. We will experimentally setup trial blends by varying relative concentrations of Zinfandel to Petite Sirah. We will look, smell, and taste to determine what minimum amount is needed to get the color darkened and still have a nice flavor and aroma.

Create tasting and testing samples of varying relative concentrations. Note the color, aroma, and taste of blends and down-select to the right combination of percentages. Provide additional columns for other parameters you may consider appropriate to your final blend criteria.

Find the blend with just the percentages of the various varietals that satisfies your color needs and still maintains the aroma and flavor profile. This can be a personal blending trial, or you can invite expert colleagues for independent checking. You may want to do this quickly or you might want to spin off a tasting social event with

Fig. 7.11 Increase the percentage of darker Petite Sirah to achieve a darker color but still retain Zinfandel taste. Keep relative concentration low enough to stay within your preferred taste and aroma profile

Table 7.10 Scale blend ratios up to production levels

Production volumes			
Blend #	Production blend volume (gal)	Production Zinfandel needed (gal)	Production Petite Sirah needed (gal)
1	600	600	0
2	600	540	60
3	600	450	150
4	600	0	600

family and friends. Fig. 7.11 shows how varying the relative concentration of Zinfandel to Petite Sirah will significantly change the color.

Table 7.10 scales up the trial concentrations to production levels for a target final volume of 600 gallons.

At this point, you either have the blending wine in your winery or contacts to find a supplier of the needed wine. You may have to pay for additional Petite Sirah, but it should be at volumes that will be a small fraction of the base volume of Zinfandel. If the blend taste trials succeed, you will still be able to bottle a high-quality blend. If it falls short of your premium wine goals, you may elect to distribute as a second label or find another buyer who is also seeking a "late in the season" blending wine.

This rational approach can be implemented quickly and provide appropriate and reliable results. After the bottling deadline is met, more time can be spent to understand the root cause and make corrections for next time. You are well down the road to improving your skills and making a higher quality wine.

Reference

Kepner, C. H., & Tregoe, B. B. (2013). *The new rational manager: An updated edition for a new world.* Princeton, NJ: Princeton Reasearch Press.

Chapter 8
Wine Quality Problem Cause Identification with the Best of the Best and Worst of the Worst (BowWow) Method

8.1 Where Does One Start? BowWow Can Help

This is a section about another powerful tool to aid in resolving winemaking problems. It is called best of best and worst of worst' or "BowWow"[1] for short. The world of winemaking is big and wonderful which unfortunately also means problems are likely and often very complex to solve. BowWow helps find plausible suspects for the cause of your problem. It looks at the link between the best and worst finished wines and the best and worst ingredients, equipment, or other vinification aspects. The simple premise is that often, the best components will make the best wine and the worst components will make the worst wine. This is a great way to start identifying good suspects. It may not isolate the true root cause of your quality problem; however, it improves your chances of identifying probable suspects.

The pyramid in Fig. 8.1 illustrates our basic nine process steps for making red wine. The quality of the grape and its growing region's sunlight and temperatures will play a big role in the quality of the finished wine (Spayd, Tarara, Mee, & Ferguson, 2002). Any one of the nine steps may maintain, improve, or worsen such quality. That first off-smell may offer a simple, straightforward clue of rotten eggs meaning too much hydrogen sulfide. It might have come from your fermentation process. Was it your yeast, or your yeast nutrients, or temperature, or what? Many problems are not obvious flaws. Many are a matter of marginal changes, out of balance conditions, or offsetting concentrations. Evaluate the history and details of your winemaking process, connect the winemaking steps to the best of best wines with no discernible rotten egg aroma and the worst of worst finished wines with the strongest off aroma. There are six major groups to check: process, materials, equipment, people, measurements, and environments amongst the best of the best and

---

[1] This process was named the BowWow method by Dr. Howard Sawhill and shared with us over a glass of wine in 2018.

© The Editor(s) (if applicable) and The Author(s), under exclusive license to
Springer Nature Switzerland AG 2020
J. Steakley, B. Steakley, *A Quest for Quality Wine, Every Time*,
https://doi.org/10.1007/978-3-030-34000-1_8

The best quality wine?

Do my quality attributes Increase or decrease?

Am I using the worst Practice ?

Store

Bottle

Clarify

Age

Press

Ferment

Crush

Harvest

Grow

Who, what, where, when, and why do my problems originate?

Am I using the best practice?

The best quality grapes?

Fig. 8.1 The world of winemaking is vast. There are many high-quality wines and there are many that are not. When you taste a glass of wine you made and it is not great, you will wonder what happened. BowWow can help. It is a method to find the Best of the best and the worst of worst. It is a powerful method that can help you focus, prioritize, and identify problem suspects

worst of the worst. These focus groups will provide us with a good of list of suspects. It covers the complete winemaking space. Look for the presence of the problem and the absence of the problem. More samples are helpful, but even a few can benefit from the BowWow method.

Let us review some of the features and benefits of the BowWow method. Table 8.1 breaks out eight positive features and associated benefits. Like other methods discussed in this book, this is a rational approach which is expected to produce a rational outcome. It is a coherent and cogent approach to establish most likely connections starting with the components of your wine and comparing that set which produces the highest quality and that set which produces the worst. Also, like other methods presented earlier, it starts with a thorough and concise problem statement. This helps frame the issue and establish details to ensure you are addressing the correct attributes of your own problem situation.

Your own historical details of your winemaking techniques provide an accurate set of facts to anchor the basis for your best quality and worst quality results. Establishing your list of common and different ingredients and equipment used can reveal items that worked and items that did not. Search for patterns or multiple occurrences of

Table 8.1 Features and benefits of the best of best and worst of worst (BowWow) method

#	Feature	Benefit
1	Rational approach	The world of winemaking is vast, complex, and full of potential problems that the BowWow approach can address. Attacks the leading issues and leads to isolating the root cause
2	Assess and clarify the problem situation	First get a good understanding of the problem and associated characteristics. Understand the details before your waste precious time and energy on solving the wrong wine problem or path to fix. This is a good start for any rational problem-solving technique
3	Collect history of process, materials, equipment, people, measurements, and environments	Establishes database of your own winemaking process and contributing factors to develop relevant candidate possibilities for the cause of your particular problem. This builds upon the detailed history of your winemaking, making it applicable and relevant
4	If the better-quality wines used a particular grape and the worst quality wines used a different grape, then that different ingredient goes on the list of suspects	This builds the tie between components of your own better-quality wines and your own lesser quality wines. It uses contrast between the high-quality and low-quality wines to bring out distinction in component effects
5	If the better-quality wines used a particular piece of equipment and the lesser quality wines use a different piece of equipment, then that goes on the suspect list	Similar to above, this builds a tie between the quality of your own wine and its components and contrasts or discriminates components of higher quality and lesser quality wines
6	Complete your list of suspects for the categories of process, material, equipment, people, measurements, and environments	This is a comprehensive assessment of your historical winemaking problem suspects built up from discriminating your own wine qualities and their respective components and winemaking steps
7	Narrow and rank order the list of suspects to well-defined cases with significant contrast in quality and understanding of grapes and vinification	This is the powerful BowWow step of finding problem suspects. Use the best of best and the worst of worst to establish the biggest contrast and most compelling basis. This will help focus the list of suspects to the leading problem suspects and provide a manageable path forward. You may end here, if the historical data is clear and science-based
8	For the top cases, there can be additional scope and effort that can confirm and clarify the root cause. This involves the design of a controlled experiment to repeat and highlight the cause of the problem	The BowWow method establishes the first step in identifying likely problem suspects. Further work is necessary to get to a confirmed root cause. You should proceed with a controlled experiment to isolate the root cause, depending upon your volume of production, cost, and schedule at risk

ingredients and equipment that most often produced higher quality. Search for patterns or occurrences of ingredients and equipment that most often produced lower quality wines. This latter set are good suspects from problems. We offer the BowWow approach as an early systematic method to identify plausible suspects for the cause of your problems. Analysis of the differences will further enable plausible science-based arguments for the best and worst quality finished wines.

Figure 8.2 illustrates the concept of separating the extremes, temporarily setting aside the ambiguous or gray areas, and narrowing the field of good problem suspects. Controlled experiments can be employed to provide additional empirical evidence of the root cause. Various problem-solving methods can and should be utilized. Each will have its own strengths and weaknesses. Your own problem situation, knowledge, and experience level may also dictate which method is appropriate to apply. Some will get you started from a clean sheet of paper and others will help carry you to the finished glass. This book is intended to teach multiple different ways and styles to logically solve a winemaking problem.

8.2 The Mystery of the Four Young Cabernet Sauvignon Wines

We all can work hard to make wine, but will it be a quality wine? Your chances will improve if you use one or more of the methods in this book. We understand these methods; however, we did not always execute properly. It not only takes understanding, but also proper execution. Let us explore another scenario to explain how to the BowWow method may be employed to identify leading problem suspects. This is what we call the mystery of the four young Cabernet Sauvignon wines.

In this scenario, we were very disappointed in a couple of vintages of our Cabernet Sauvignon. We were frustrated. Sometimes life intervenes. It just so happens that we were moving our entire winery operation to another part of town. That is big disrupter, however, it is really not a good excuse to make poor quality

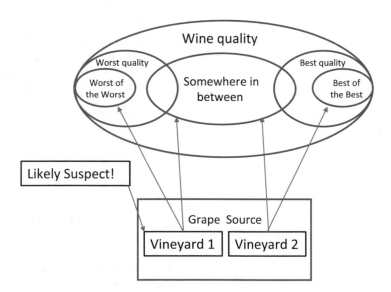

Fig. 8.2 BowWow identifies extremes, separates the ambiguous, and helps capture suspects

wine. Be wary of ostensible excuses that may in fact not be the root cause of your problem. After a brief soul-searching spell, we went back to basics. We have four well-intentioned but varying quality Cabernet Sauvignon wines in this scenario. Let us use the BowWow method to identify likely poor quality suspects.

Our problem begins with the 2012 vintage. The color has a brown tint and a slight sherry or nutty aroma. It is from a nice vineyard in a nice growing region and the harvest numbers look pretty good. However, the aroma and taste are unappealing. We changed vineyards, but still in a nice grape growing AVA. At first glance, maybe it was grapes from that vineyard. The 2013 was not much better. It started with similar sugar and acid, but lower pH. It smelled and tasted quite different, but still was not appealing. 2014 took another notch back down the quality ladder. It looked slightly browner, had some medicinal notes, and was even less appealing. We needed to pay closer attention to our winemaking practice and analysis methods. It was not until the 2015 vintage that we realize a discernible increase in the quality of the wine.

We recommend a thorough review of your own notes.[2] By keeping logbooks, historical records, and pictures of your grapes and vinification process, you will have captured clues to the cause of your problem. Table 8.2 shows the four different vintages of Cabernet Sauvignon. These wines were of varying quality. Even though the 2013 growing season was called "near epic" and 2014 also good, our finished wines did not fare so well. The quality scores are examples from independent wine critics in local wine competitions. They were all tasted about 3 years after the vintage. Not very good scores, however, they did improve in 2015. The 2014 was the worst and the 2015 was the best. Why was this?

In Fig. 8.3, the right six columns are the top-level categories we use throughout this book and discuss in detail in earlier sections. We made our own detailed self-assessment of the quality of our process and conditions for each vintage. We grouped the details into these six categories which represent our view of the entire universe of possibilities. We took some liberties with the "people" category given ourselves the improving score of 3 to 2 with practice with 2 years vintages. For this example, consider this a biased self-assessment and do not forget to always learn and train your team. Consider the other categories and look for the best practices, the worst practices, and associate patterns with the best wine and the worst wine.

List the details of your own process, materials, equipment, people, measurements, and environments and associate with the corresponding finished wine. At first, every input or condition that went into the worst wine is a concern. Also note that every input or condition that went into the best wine is not necessarily exonerated or cleared. Multiple other conditions may have been so overwhelmingly good that they made up for problems or risks in other categories. We do not have a clear science-based path at this point, but we will detect emerging patterns. We are now looking for suspects that we will investigate further. To narrow the list of suspects, look for the biggest contrast or gaps. Review the very best of the best and the very

---

[2] Refer to the basic winemaking Chap. 4 regarding important record keeping throughout the various phases. These notes are invaluable to retracing the activities and results of prior seasons. They will contain the clues to resolving your problems.

Table 8.2 The problem row shows sensory observations. The other rows show source and chemistry data. The critic score row shows lower-quality scores for earlier vintages with some improvement in 2015

Varietal	Cabernet Sauvignon	Cabernet Sauvignon	Cabernet Sauvignon	Cabernet Sauvignon
Vintage	2012	2013	2014	2015
Problem observation	A little brown with faint sherry like or oxidized odor	Dark black color, "pruney," off nose, maybe VA	Brown, oxidized color, aroma of vinegar, slight medicinal	Nice clear deep purple color, hint of vegetal, a bit tart, slightly astringent
Grape source	Vineyard 2	Vineyard 1	Vineyard 1	Vineyard 1
Sugar at harvest	26° brix	26° brix	24.5° brix	22.5° brix
TA at harvest	6.2 g/L	6.2 g/L	7.2 g/L	7.7 g/L
pH	3.5	3.0	3.0	3.2
VA	Not measured	0.1	0.11	0.1
Critic score (20max)	10	11	10	14

			Vinification practice assessment (best = 1, worst=5)					
Wine Varietal	Wine Vintage	Finished wine quality score (0-20)	process	materials	equipment	people	measurements	environment
Cabernet Sauvignon	2012	11	4	4	3	3	4	3
Cabernet Sauvignon	2013	11	4	3	3	3	3	2
Cabernet Sauvignon	2014	10	5	3	2	2	5	3
Cabernet Sauvignon	2015	14	2	2	2	2	2	3

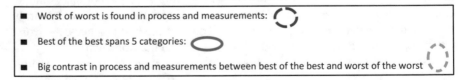

- Worst of worst is found in process and measurements:
- Best of the best spans 5 categories:
- Big contrast in process and measurements between best of the best and worst of the worst

Fig. 8.3 Research and mine your own history of your grapes and vinification quality. Look for best of best and worst of worst practice. Evaluate links and biggest contrast for problem suspects

worst of the worst. Our very best of the best wine is the 2015 and the very worst of the worst is the 2014, so we will focus on these to find our problem suspects.

Figure 8.3 showed us the biggest contrast or gap in the quality of winemaking was in process and measurements. Detailed review of the processes showed significant differences in best practice in the frequency or number of wine barrel top-offs during aging. Table 8.3 breaks out contrast in quality of aging practice and possible problem suspects. The risk and consequences of infrequent oak barrel top-off are the development of unwanted oxidation in the headspace and causing browning and quality impacts (Boulton, Singleton, Bisson, & Kunkee, 2010). There were also more frequent sensory measurements (visual, smell, and taste) of the wine itself (Peynaud, 1983). This is a probable suspect for the cause of our low-quality wine. Note how there are differences in the other categories, however, we are searching for the largest or most extreme differences. Other categories varied but not with extreme differences. Yes, the vintages had varying climates and coming off drought years, however, drip irrigation was used in all 4 years.

The measurements category includes testing wine chemistry as well as sensory measurements. What do the wines look, smell, and taste like (Jackson, 2009). This includes the frequency of which these measurements or tasting were made. Tasting more frequently will detect emerging problems before they grow to big issues. In 2014, our records show we did not inspect, nor taste, nor measure the wine more than once or twice during the year. Ugh! We were trying to learn and get better and at least for 2015, we inspected and tasted the aging wine almost monthly. Table 8.4 breaks out measurement details of the best of the best (2015) and worst of the worst (2014). Frequent inspection and sensory testing of the wine during aging was a big contrast between 2015 and 2014. We tried to discipline ourselves to top-off almost every month and include a careful tasting. These checks will have revealed a degrading wine and the issue of air in the headspace. The measurements would not in themselves prevent the problem, but they indicate conditions and emerging issues. Conducting sensory tests occasionally is better than never, however, very few sensory tests is another strong problem suspect. We have found a prime suspect for our low-quality 2014 wine.

8.3   The Mystery of the Five Meritage Blends

In this scenario, we start with a very high-quality Meritage blend in 2010. It tasted great. We liked it. The critics like it. We thought we had this winemaking skill all locked up. We thought next year's vintage would be even better. We thought we

Table 8.3 These aging process details show big contrast in top-off frequency between best of best and worst of worst wine. BowWow indicates this is likely suspect and should be pursued as possible problem cause

Wine	Process— aging phase	Possible best or worst practice	Contrast between best of best and worst of worst	Plausible explanation of problem	Possible problem suspect
Best of best: 2015 Cabernet Sauvignon	Aging	Best: Frequent top-off of wine barrels and tasting; near monthly	Top-off wine barrels 10–12 times per year limits headspace oxygen exposure	Air caused oxidation and browning of wine	Infrequent top-off of wine barrels wine during aging
Worst of worst: 2014 Cabernet Sauvignon	Aging	Worst: Only twice in 12 months. Lack of attention to process in winery	Top-off 2× per year allowed oxidation of the wine		

Table 8.4 Frequency of measurements or sensory tests during aging shows prominent gap in quality of practice between best of best wine and worst of worst wine. Another good suspect

Wine	Measurements—aging	Possible Best or worst practice	Contrast between best of best and worst of worst	Plausible explanation of problem	Possible problem suspect
Best of best: 2015 Cabernet Sauvignon	Sensory and chemistry tests during aging	Best: Inspected, smelled, and tasted almost monthly	Big difference in tasting frequency. Measured and tasted 5–10 times as often during aging	Rarely tasting failed to reveal degrading wine. Air caused oxidation and browning	Infrequent sensory testing of wine during aging
Worst of worst: 2014 Cabernet Sauvignon	Sensory and chemistry tests during aging	Worst: Inspected and tasted once or twice during a year period			

could go on autopilot and pump out case after case of great wine. Yikes, we could not have been more wrong! Each year the wine critics dropped the quality scores of our wine. The comments were getting more pointed. How could this be? Let us figure this out. We have five vintages of various Meritage blends. We have also assembled a large data set contained in Tables 8.5, 8.6, 8.7, 8.8, 8.9, 8.10, and 8.11 showing wine critic comments, scores, harvested grape chemistry, and grape source AVAs. These may participate in quality issues within our process (grape growing or harvest), materials (the grape quality and or varietal), or measurements (juice or wine chemistry) groups. Whether you are following the details on an on-going basis, or you get a big data set dropped on your lap, BowWow methods can help triage the situation before you jump into costly randomized controlled trials. You should be able to sort out the best of best and worst of worst within the ingredients of your best and worst quality wine.

Of our five red wine Meritage blends, the 2010 vintage was the highest quality. This is the first blend in our series of Cabernet or Merlot dominant Bordeaux-style blends. The following data includes the quality level as determined by wine competition blind tasting scores. The scores are an average of 3 or 4 wine judges at county fair wine competitions. They used a 20-point scale similar to what was discussed in the quality Chap. 2. There is some scoring and descriptive variation amongst the judges for the same wine, however, we mostly agreed with their assessments. As we already mentioned, the quality of subsequent vintages seemed to go downhill or not improve for the next couple of years. All the time, we thought we were getting smarter, but we were not really paying attention to winemaking details. Alas, there was a discernible quality uptick in 2015 and 2016. We will focus on grape growing region, relative grape varietal concentration, and grape chemistry at harvest and see what BowWow tells us. Tables 8.5, 8.6, 8.7, 8.8, 8.9, 8.10, and 8.11 contain data on the average score, blend grape concentrations, and grape chemistry (Iland, Bruer, Edwards, Weeks, & Wilkes, 2004). Table 8.5 contains quality comments and scores from the judges for each of the five blends. Tables 8.6, 8.7, 8.8, 8.9, 8.10, and 8.11 provide variety concentrations, grape source, chemistry, and aging information for each grape varietal in our blends.

Table 8.5 Extensive data available on grape ingredients and blend ratios for our mysterious Meritage blend vintages. Let us use BowWow techniques to find problem suspects

Varietal	Meritage blend	Meritage blend	Meritage blend	Meritage blend	Meritage blend
Vintage	2010	2011	2012	2014	2016
Expert critic comments	Good color, fruity, appropriate, taste currants, lasting finish	Some murkiness, small nose, some chemical, some fruit but watery, not bad finish	Ok appropriate, slightly brown color, jammy, cooked, raisiny, slightly tart, nice wine if slightly stewed, not fresh	Dark, good color extraction, menthol, stewed, cooked, vegetal, canned prunes, vinegar, complex but in a weird way	Deep dark ruby red, fruity delicate nose, black fruits, slightly tart, balanced, thin, light body, fine character
Meritage blend critic scores average (20max)	18	15	14	13	16

Table 8.6 Cabernet Sauvignon concentrations, source data, chemistry, and aging information

Grape ingredients and aging data: Cabernet Sauvignon					
Vintage	2010	2011	2012	2014	2016
CS grape concentration	50%	33%	63%	72%	62%
CS grape source	Vineyard 3. Mendocino AVA	Vineyard 2: Santa Cruz Mountain AVA	Vineyard 3. Mendocino AVA	Vineyard 1: Santa Cruz Mountain AVA, sunny blocks	Vineyard 1: Santa Cruz Mountain AVA, sunny blocks
CS sugar at harvest	25° brix	26° brix	26° brix	25° brix	25° brix
CS pH	3.5	3.6	3.5	3.5	3.5
CS TA at Harvest	Not measured	5.6 g/L	6.2 g/L	8.9 g/L	6.4 g/L
CS volatile acidity	Not measured	Not measured	Not measured	0.1 g/L	0.1 g/L
Aging container	Oak	Oak	Oak	Stainless and oak	Oak
New or used barrel	Used	New	Used	New	New
Aging time	18 months	23 months	24 months	33 months	21 months
Aging top-off	Near monthly	Every 4 months	Every 6 months	First in 9 months, second in 12 months, then every 6 months[a]	Monthly

[a]Moved Wine, Winery, and personal residence. Lost attention on wine and winery and risks up

Table 8.7 Merlot concentrations, source data, chemistry, and aging information

Grape ingredients and aging data: Merlot					
Vintage	2010	2011	2012	2014	2016
Merlot grape concentration	50%	60%	35%	0%	15%
Merlot grape source	Vineyard 3: Mendocino AVA	Vineyard 2: Santa Cruz Mountain AVA	Vineyard 2: Santa Cruz Mountain AVA	NA	Vineyard 1: Santa Cruz Mountain AVA, shady block
Merlot sugar at harvest	26° brix	23° brix	25° brix	25° brix	23° brix
Merlot TA at harvest	Not measured	6.9 g/L	6.2 g/L	8.4 g/L	6.2 g/L
Merlot pH	3.5	3.5	3.9	3.6	3.5
Merlot VA	Not measured	Not measured	Not measured	0.12 g/L	0.1 g/L
Aging container	Oak	Oak	Oak	NA[a]	Oak
New or used barrel	Used	New	Used	New	New
Aging time	18 months	23 months	24 months	NA	21 months
Aging top-off	Near monthly	Every 4 months	Every 6 months	NA	Monthly

[a]Tossed Merlot in 2016 (sad day) very poor quality, smelled medicinal, oxidation, just not good

Table 8.8 Cabernet Franc concentrations, source data, chemistry, and aging information

Grape ingredients and aging data: Cabernet Franc					
Vintage	2010	2011	2012	2014	2016
CF grape concentration	0%	1.0%	0%	15%	10%
CF grape source	NA	Vineyard 2: Santa Cruz Mountain AVA	NA	Vineyard 1: Santa Cruz Mountain AVA, sunny block	Vineyard 1: Santa Cruz Mountain AVA, sunny block
CF sugar at harvest	NA	23° brix	NA	25° brix	23° brix
CF TA at harvest	NA	9.3 g/L	NA	8.4 g/L	6.2 g/L
CF pH	NA	3.5	NA	3.6	3.5
CF VA	NA	Not measured	NA	0.12 g/L	0.1 g/L
Aging container	Oak	Oak	NA	Stainless and oak	Oak
New or used barrel	Used	New	Used	New	New
Aging time	18 months	23 months	NA	CF combined with CS in 2015	21 months
Aging top-off	Near monthly	Every 4 months	NA	First in 9 months, second in 12 months, then every 6 months	Monthly

Table 8.9 Petit Verdot concentrations, source data, chemistry, and aging information

Grape ingredients and aging data: Petit Verdot					
Vintage	2010	2011	2012	2014	2016
PV grape concentration	0%	6.0%	2.0%	9%	10%
PV grape source	NA	Vineyard 2: Santa Cruz Mountain AVA	Vineyard 2: Santa Cruz Mountain AVA	Vineyard 1: Santa Cruz Mountain AVA, sunny block	Vineyard 1: Santa Cruz Mountain AVA, sunny block
PV sugar at harvest	NA	Not measured. Field blend	Not measured. Field blend	23° brix	22° brix
PV TA at harvest	NA	Not measured. Field blend	Not measured. Field blend	6.4 g/L	8.7 g/L
PV pH	NA	Not measured. Field blend	Not measured. Field blend	3.6	3.1
PV VA	NA	Not measured. Field blend	Not measured. Field blend	0,1 g/L	0.1 g/L
Aging container	Oak	Oak	Oak	Stainless and oak	Oak
New or used barrel	Used	New	Used	New	New
Aging time	18 months	23 months	23 months	PV combined with CS in 2015	21 months
Aging top-off	Near monthly	Every 4 months	Every 6 months	First in 9 months, second in 12 months, then every 6 months	Monthly

Figure 8.4 maps the grape source growing region and specific vineyard to wine quality. The first wine is made with wines from the Mendocino AVA and the other four are made from grapes grown within the Santa Cruz Mountain AVA (Ray & Marinacci, 2002). The Mendocino AVA is known for Zinfandel and Mediterranean red varieties such as such as Syrah, Petit Syrah, Carignane. However, it is also home to very nice vineyards that grow Cabernet Sauvignon and Merlot. Our 2010 blend was very nice. In subsequent years, we generally thought going with prime Santa Cruz Mountain growing regions, our wine will be even better! The Santa Cruz Mountain AVA is known for its mountain topography which means there are different sun angles, temperatures, soils, etc. There are many high-quality wines made from grapes grown in this AVA, including Cabernet Sauvignon, Zinfandel, Cabernet Franc, Malbec, and others (Sullivan, 1982). For our preferred style, this sure sounds like a good fit. The mountainous widely varying vineyards and small blocks may or may not provide the optimal growing conditions. Pay attention to the actual vineyard, its layout, its canopy management (Smart & Robinson, 1991), and the quality of the grapes you end up using in your wine.

Table 8.10 Malbec concentrations, source data, chemistry, and aging information

Grape ingredients and aging data: Malbec

Vintage	2010	2011	2012	2014	2016
Malbec grape concentration	0%	0%	0%	1%	2%
Malbec grape source	NA	NA	NA	Vineyard 1: Santa Cruz Mountain AVA, partial sun block	Vineyard 1: Santa Cruz Mountain AVA, partial sun block
Malbec sugar at harvest	NA	NA	NA	24° brix	Not measured. Field blend
Malbec TA at harvest	NA	NA	NA	8.5 g/L	Not measured. Field blend
Malbec pH	NA	NA	NA	3.4	Not measured. Field blend
Malbec VA	NA	NA	NA	0,1 g/L	Not measured. Field blend
Aging container	NA	NA	NA	Stainless and oak	Oak
New or used barrel	Used	New	Used	New	New
Aging time	NA	NA	NA	Malbec combined with CS in 2015	21 months
Aging top-off	NA	NA	NA	First in 9 months, second in 12 months, then every 6 months	Monthly

Table 8.11 Alicante Bouschet concentrations, source data, chemistry, and aging information

Grape ingredients and aging data: Alicante Bouschet

Vintage	2010	2011	2012	2014	2016
AB grape concentration	0%	0%	0%	3%	1%
AB grape source	NA	NA	NA	Vineyard 1: Santa Cruz Mountain AVA, partial sun block	Vineyard 1: Santa Cruz Mountain AVA, partial sun block
AB sugar at harvest	NA	NA	NA	Not measured. Field blend	Not measured. Field blend
AB TA at harvest	NA	NA	NA	Not measured. Field blend	Not measured. Field blend
AB pH	NA	NA	NA	Not measured. Field blend	Not measured. Field blend
AB VA	NA	NA	NA	Not measured. Field blend	Not measured. Field blend
Aging container	NA	NA	NA	Stainless and oak	Oak
New or used barrel	Used	New	Used	New	New
Aging time	NA	NA	NA	33 months	21 months
Aging top-off	NA	NA	NA	First in 9 months, second in 12 months, then every 6 months	Monthly

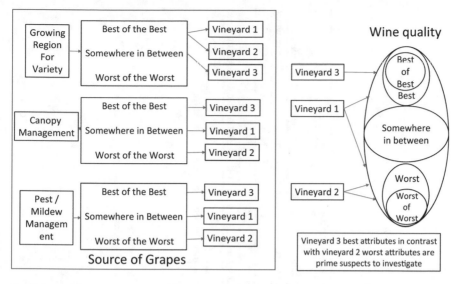

Fig. 8.4 Our Grapes came from different vineyards. Check the best and worst features applicable to each vineyard for further clues to problem suspects. Prioritize your investigation on the best and worst aspects before you spend a lot of time on things "somewhere in between"

Initial problem solving can be overwhelming when starting with a low-quality wine and clean sheet of paper. It can also be overwhelming when you have got lots of data to review from multiple years, types, and winemaking histories. Save time with initial search for problem suspects by eliminating wines that have quality scores somewhere in between the best and worst (reference Table 8.12). You can always come back to these later, if your hunt for problems does not produce a viable suspect or further controlled experiments do not find the root cause.

Let us take a deeper dive. What are the typical temperatures and rainfall for these two grape vineyards and growing regions? We still have the six categories of the best and worst wine to evaluate.

Our first group is process. This group has nine basic process steps to make wine. For this scenario, the steps or phases where there is not much difference between the best and worst wine are shown as "N/A" in Table 8.13. There might be differences in the ingredients. The quality of the grapes themselves could be drivers. This can be a result of the grape growing methods and or harvest timing. In addition, the aging duration and top-off schedule was significantly differently between the best and worst vintages. Check your conditions, however, for this scenario we have crossed out six phases and are considering three; grow, harvest, and age.

We have narrowed the process phases to grow, harvest, and age by assessing significant differences in best or worst practice between our best 2010 Meritage blend and our worst, the 2014 blend.

Table 8.12 Start narrowing universe of possibilities with BowWow considerations

Wine quality				Grape growing and vinification assessment template					
Name	Vintage	Blend	Quality score (0–20)	Process	Materials	Equipment	People	Measurement	Environment
Hawk Tail Red	2010	1[a]	18	Best or worst	Best or worst	Best or worst	Best or worst	Best or worst	Best or worst
Hawk Tail Red	2011	2[b]	15	N/A	N/A	N/A	N/A	N/A	N/A
Hawk Tail Red	2012	3[c]	14	N/A	N/A	N/A	N/A	N/A	N/A
Hawk Tail Red	2014	4[d]	13	Best or worst	Best or worst	Best or worst	Best or worst	Best or worst	Best or worst
Hawk Tail Red	2016	5[e]	16	N/A	N/A	N/A	N/A	N/A	N/A

Eliminate initial problem search space by not considering wines with quality characteristics "somewhere in between" best and worst (noted as N/A). Dive deeper into the two remaining for now

[a]Blend 1: 50% Cabernet Sauvignon, 50% Merlot
[b]Blend 2: 33% Cabernet Sauvignon, 60% Merlot, 6% Petit Verdot, 1% Cabernet Franc
[c]Blend 3: 63% Cabernet Sauvignon, 35% Merlot, 2% Petit Verdot
[d]Blend 4: 72% Cabernet Sauvignon, 15% Cabernet Franc, 9% Petit Verdot, 1% Malbec, 3% Alicante Bouschet
[e]Blend 5: 62% Cabernet Sauvignon, 15% Merlot, 10% Cabernet Franc, 10% Petit Verdot, 2% Malbec, 1% Alicante Bouschet

Table 8.13 Evaluate the differences in the nine process phases for your best and worst wine

Wine quality				Grape growing and vinification assessment template							
Name	Vintage	Blend	Quality score (0–20)	Grow	Harvest	Destem crush	Ferment	Press	Age	Clarify	Bottle	Store
Hawk Tail Red	2010	1	18	Best or worst	Best or worst	N/A	N/A	N/A	Best or worst	N/A	N/A	N/A
Hawk Tail Red	2011	2	15	Best or worst	Best or worst	N/A	N/A	N/A	Best or worst	N/A	N/A	N/A
Hawk Tail Red	2012	3	14	Best or worst	Best or worst	N/A	N/A	N/A	Best or worst	N/A	N/A	N/A
Hawk Tail Red	2014	4	13	Best or worst	Best or worst	N/A	N/A	N/A	Best or worst	N/A	N/A	N/A
Hawk Tail Red	2016	5	16	Best or worst	Best or worst	N/A	N/A	N/A	Best or worst	N/A	N/A	N/A

In this scenario, there are potential differences in the phases grow, harvest, and age (the others are indicated as N/A)

8.3.1   BowWow Aids in Reducing Search Space within Process Phases

The growing conditions depend upon the region and its temperatures, rain, soil, and pests. Table 8.14 shows similar average annual temperatures at the vineyards that sourced the grapes that made the best and worst wines. Table 8.14 also shows the average rainfall was much heavier in vineyard two. Remember these are annual averages.

Annual averages provide a broad indicator; however, they do not reveal monthly changes. Monthly conditions can and will vary significantly throughout the growing season. Month to month changes in temperature and rainfall can be big drivers in the grapes ripening properly. The average rainfall was heavier in vineyard 3 which produced the grapes used in the best wine. However, rain levels will change year to year and month to month. Some of this can be compensated for with managed irrigation. In heavy drought years there are limits to compensating for water shortfalls.

Let us look at monthly averages as shown in Fig. 8.5 and compare these two growing regions. The monthly climatic average plot shows monthly climate averages over a few decades. Note there are mid-day temperature differences in the late summer and early fall. These differences may be significant or not. However, they should be reviewed for the problem suspect list. These kind of temperature differences will impact berry growth and ripening and will impact the length of the growing season. This may have bigger or smaller impacts depending on your grape variety.

Each year will have different weather and or pest conditions. These will drive different vineyard management methods; however, extremes may not be able to be corrected. Let us look closer at wine grape growing conditions and or issues for our best year, 2010 and for our worst year 2014. Details of wine grape weather events are shown in Table 8.15. Note how vineyard 1 and 3 have different characteristics, however, minor differences can be mitigated with proper vineyard management and harvest methods.

It is also important to check your data to build confirmation or uncover ostensible inconsistencies. How did the growing conditions compare with the actual wine chemistry of the harvested grapes? Are they consistent? Can inconsistencies be explained by compensating vineyard management? If it was too hot or if the grapes were too long on the vine, the acid might be too low (Mullins, Bouquet, & Williams, 1992).

We aged both the best and worst vintages in oak barrels, however, there were some differences. Figures 8.6, 8.7, and 8.8 show detailed equipment logs with information on aging time and barrel differences between the best and worst wines. Note how useful it is to have readable logbooks one can refer back to in order to sort out what really happened! The worst wine was aged after blending all the varieties together in one new French oak barrel for 2 years. This is a plausible process; however, this could impart a huge amount of oak. Keep it on the problem suspect list.

Table 8.14 Our best and worst wines are made with grapes from two different growing regions. These are annual average temperatures and annual average rainfall

Item	Vineyard 1	Vineyard 3	Comment
Annual average high temp	71.7 °F	72.4 °F	Not significant, however, check next level of detail of monthly averages
Annual average low temp	46.9 °F	45.6 °F	Not significant, however, check next level of detail of monthly averages
Annual average rainfall	24 in.	40 in.	Significant. However, this might be normalized by irrigation

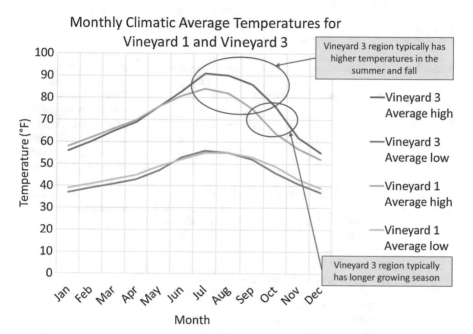

Fig. 8.5 Monthly climatic averages will reveal typical differences throughout a growing season where annual averages do not. The best and worst wines were made from grapes from these two different vineyards

The aging duration and top-off frequency standout as significantly different between the best and worst wine. The worst wine was aged for 33 months as compared to 18 months for the best wine. The 33 months is the total of the time in a stainless tank after harvest plus this time in an oak barrel. This is quite a bit longer. This in itself may not be a problem; however, the long duration adds risk of oxidation if the oak barrels are not topped off regularly. This should go up near the top of our problem suspects. If it were done properly and the desired style was very oaky with notes of vanilla, then this might not be a problem, however, this is very long compared to our best wine and very long compared to most quality wine standards. There are many risks with long oak barrel aging, including risk of oxidation, browning, and spoilage of the wine. (Flor et al., 2018) The other red flag in aging is the very dramatic top-off

Table 8.15 Look to the weather and growing season details for the vintages of your best wine and your worst wine. Look at characteristics that are above or below expected standards

Item	Harvest year weather for best wine	Harvest year weather worst wine	Comment
Vintage	2010	2014	Every vintage potentially has different conditions
Vineyard	• Vineyard 3: Mendocino AVA	• Vineyard 1: Santa Cruz Mountain AVA	Different weather affects different AVAs in different ways
Pre-season	• A cold, wet spring	• Half annual rainfall prior to budbreak	More early rain in 2010 than in 2014
Initial season	• Late budbreak	• Early budbreak	Timing of budbreak may or may not be significant
Mid-season	• Generally cool growing season • Heat spike hit in August	• Warmer than usual nights mid-season	2010 vineyard did not experience sun damage to crop as was the case in some neighboring AVAs
End of season	• Harvested late but just in time before end of October rains	• Early harvest • Drought year reduced yield	Yields within 2010 Mendocino AVA were good and growing season length allowed grapes to ripen
Canopy management	• Less morning fog than usual • Morning temperatures cool but not too cold • Did not have to pull leaves to open the vines to the sun	• Canopy management was reasonably normal, but sporadic	Some differences, but not extreme
General growing season	• A wet spring followed by a late start and long, cool growing season ending with a heat spike allowed grapes to reach maturity	• Drought year reduced yield	Duration from veraison to harvest in 2014 was similar to past years

frequency difference between the worst and best wine. The best wine was almost monthly and the worst wine seemed to have only been done four or five time in 33 months. This had long spans of 9–12 months with no barrel top-off. Oak barrels allow evaporation and the increasing headspace within the barrel fills with air. This aging process and risk of oxidation goes to the top of our problem suspect list.

8.3.2   BowWow Revelations on Materials

We are using various measurements to indicate the quality of our grapes, the primary material ingredient in our wine. The data table shows essential grape juice chemistry at the time of harvest. This includes sugar, Titratable Acidity (TA), pH,

Fig. 8.6 Equipment log documents 2010 Ukiah Merlot time in used barrel. This level of detail is necessary to understand history of the Merlot component of this blend

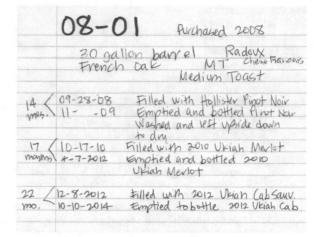

Fig. 8.7 Equipment log 08-02 documents the Ukiah Cabernet Sauvignon component time in its new oak barrel. Knowledge of aging attributes of both components of blends is key to search for suspects

and volatile acidity. We added VA measurements in the later years because of a concern of volatile acidity showing up in the earlier wines as well as better availability of VA measurements. The Cabernet Sauvignon sugars are all reasonably high in the range of 25–26° brix. The Merlot grapes are a few degrees lower. This may or may not be consistent with your preferred style. If the harvest date is late in the season it may be an indication of the Merlot not quite ripening as much as desired. A reasonable pH is in the range of 3.3–3.6 and the Cabernet Sauvignon grapes all fall within this range. The Merlot grapes also are within the nominal range, except the 2014 which is up at 3.9. The TA should be within a range of 6–8 g/L. The 2014 Cabernet Sauvignon TA is too high and the 2011 is low, but the 2016 fell within range. The 2014 Merlot TA is slightly too high.

Fig. 8.8 Equipment log 14-02 indicates aging time and barrel for our worst wine. Compare these characteristics with aging attributes from the above best wine, the 2010 vintage. Note that the logbook entry shows this was a new oak barrel and thus was not a likely suspect

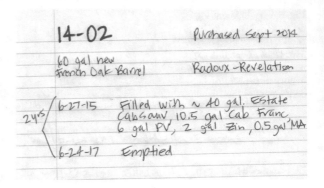

Check the other grape varieties and chemistry for nominal or out of range conditions. Juice and wine chemistry measurements are key indicators. Wine chemistry levels should be considered thresholds of acceptability. Our pH measurements show 3.5, which is good being between 3.3 and 3.6, for both the Cabernet Sauvignon and Merlot in the best wine from vineyard 2. The worst wine, 2014, showed a pH of 3.6, on the edge of a reasonable range, for the Merlot from vineyard 1. This could be a contributing factor to the problem, but it is not extreme.

8.3.3   BowWow Revelations on Equipment

The equipment was compared between that used in the best (2010) and worst (2014) vintage blends. The data in the equipment logs show the 2010 as aged in used French oak vs. new oak in 2014. We used a new medium toast French oak barrel and the best wine was aged in used French oak barrel. The used barrel is a more neutral barrel having been used in a prior season. It still could impart oak characteristics, but much less for a given time period. New oak barrels impart more significant woody vanilla flavors into the wine. The duration of the aging in the new oak barrels used in the worst wine was almost twice that of the duration in the best wine. The additional woody effect of using new barrels vs. used combined with the twice the duration significantly increases the impact. This is a notable difference and should go on our list of possible suspects. Other equipment used throughout the nine basic winemaking phases was the same.

8.3.4   BowWow Revelations on People

Did our vineyard and winery crew change between the best and worst wines? Both Joyce and I led all activities. We did get help and support during the peak harvest and bottling events. Our experience and knowledge improved every year. In this sense, our expertise should be scored higher. However, during the aging process phase, we were distracted with moving the winery, and our home residence. These were significant efforts that involved a lot of time and energy. They were a big distraction to grape growing and winemaking. This lack of attention to detail is a strong factor in the root cause of why we failed to top-off the 2014 wine properly. We had a very good understanding of the need for managing the barrel aging and the need for frequent topping, however, other important life activities outside of winemaking took a higher priority. This significantly reduced the execution and quality of our process and measurements. This should be appended to our wine aging problems as possible root cause suspects.

8.3.5   BowWow Revelations on Measurements

Our measurements were very much the same between the 2010 and 2014 wines, however, the timing and frequency of measurements can be just as critical as the type of measurements. Section 8.3.2 discussed materials via the various measurements we made. The time and frequency of our checking the wine during aging was very bad for the worst 2014 vintage. One should inspect and taste your wine frequently during the aging process. Once in 9 months is very poor and in contrast to monthly of the best wine is very significant. This could even be called shameful. If you want to barrel age high-quality wine, think about checking it once or twice a month. The measurement aspect of managing barrel aging is important. The poor measurement aspect of the 2014 low-quality wine should go on the problem suspect list.

8.3.6   BowWow Revelations on Environments

The environment includes the weather during grape growing and harvest. It also includes the temperature of fermentation, aging, and bottle storage. These all have a significant impact on wine quality. For this particular scenario of the five wines, the grape growing regions were different and the weather, including rain and water in the ground were different; however, the vineyard management was adequate and the grape chemistry revealed that either crop was suitable for making a quality wine. Environmental conditions were accommodated making this particular aspect equal in terms of wine quality impact.

8.3.7   Assessment for the Best of the Best and Worst of the Worst of Our Five-Blend Mystery

BowWow has narrowed the universe of an extremely large number of problem causing issues to a manageable set associated with the best of the best wine and worst of the worst wine. We have further considered the six possible groups of process, materials, equipment, people, measurements, and environments. We have assigned numerical scores in accordance with the quality of conditions and quality of practice to each as shown in Fig. 8.9.

In this assessment, the highest quality was assigned a score of one and the lowest quality received a score of five. The lowest quality practice was in the process category for the worst wine. This received a five and was assessed for the low-quality of aging process in barrels in which the suspect barrel was only topped off once during a nine-month period.

We assigned the 2014 wine, the worst wine, a five for its process because of the extreme worst of worst practice for not topping off the wine a barrel for over 9 months. The best wine received a three for process with near monthly wine top-off, but still could improve with more frequent top-off and testing. The grape materials and yeast were sufficiently good quality by virtue of chemical test results and grape sensory inspections, that we gave both a three. They both could improve with additional visual and taste testing of grapes before harvesting, additional acid tests, and results within a better range. The equipment underwent minor improvements and better cleaning preparation in the 2014, so we went from a three to a two. Our

Wine Quality				Grapes and Vinification practice relative assessment (best = 1, worst=5)					
Wine	Vintage	Varietal Blend	Wine quality score (0-20)	process, aging phase	material	equipment	people	measure-ments, timing and sensory	environ-ment
Hawk Tail red	2010	50% Cabernet Sauvignon 50% Merlot	18	3	2	3	3	2	3
Hawk Tail red	2014	72% Cabernet Sauvignon 15% Cabernet Franc 9% Petit Verdot 1% Malbec 3% Alicante Bouschet	13	5	3	2	2	4	3

- Worst of worst (approximate) is found in process , the aging phase and measurements :

- Best of the best (approximate) spans all 6 categories:

- Big contrast in process and measurements between best of the best and worst of the worst

Fig. 8.9 Note the better quality of practice scores for the best wine and the lower quality of practice scores for the worst wine. There are significant quality differences for aging process and aging measurement timing and type. These are likely problem suspects

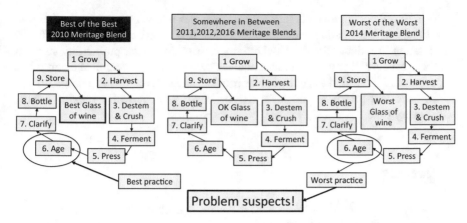

Fig. 8.10 BowWow methods narrow the universe of possibilities and isolate problem suspects. Attributes of the best of the best contrast with the worst of the worst

training and knowledge improved between 2010 and 2014 and therefore we assigned a three to 2010 and a two to 2014.[3] For the measurements group we assigned a two to the 2010 wine, however, because of the 2014 poor quality measurements during aging we assigned a four. The environments were different but acceptable, so we assigned a three to both the best and worst vintages. Given our relative scores between the worst of the worst and best of the best, it is now clear that the aging process and timing of measurements during aging are prime problem suspects for our poor quality 2014 blend.

8.4   BowWow Summary

This BowWow process is used for early problem solving to identify problem cause suspects. Figure 8.10 illustrates the extreme best and worst winemaking practice within the best and worst of this scenario. This is particularly helpful when all you have is a bad glass of wine, and a file full of notes. One can temporarily set aside all that ambiguous data that lies "somewhere in between." There is a big wide world of possibilities. We have presented two examples to show how to start and work the BowWow process. It does take time and energy but is also efficient and effective at getting you focused on the most likely problem causes. Identify your best of your best wine and the worst of your worst wine. Review the quality of your process, materials, equipment, people, measurements, and environments. Compare and contrast elements of the best with the worst. Begin your list of problem suspects with the worst quality elements of your worst wine. Compare and contrast with the same aspects, but better executed elements of your best wines.

---

[3] Note how even with better training and experience, if your people are not paying attention to the wine and not executing well, you can still end up with bad results as we did in 2014.

This BowWow approach is rational and provides a logical way to organize your vast library of information and supporting data. It is an effective first filter of your information. It reduces the time and energy researching lower probability suspects for the root cause of your poor quality wine. There is further research and controlled experimentation needed to get to a clear root cause, however, this is a powerful tool to organize and prioritize your data as you begin your quest for making quality wine, every time.

References

Boulton, R. B., Singleton, V. L., Bisson, L. F., & Kunkee, R. E. (2010). *Principles and practices of winemaking*. Davis, CA: Springer.

Flor, F. J., Leiva, F. J., Garcia, J. L., Martinez, E., Jimenez, E., & Blanco, J. (2018). Environmental impact of wine aging process in oak barrels in wineries of La Rioja (Spain). *American Journal of Enology and Viticulture, 69*(3), 302–306.

Iland, P., Bruer, N., Edwards, G., Weeks, S., & Wilkes, E. (2004). *Chemical analysis of grapes and wine: Techniques and concepts*. Adelaide: Patrick Ilan Wine Promotions PTY LTD.

Jackson, R. S. (2009). *Wine tasting: A professional handbook*. Ontario: Elsevier.

Mullins, M. G., Bouquet, A., & Williams, L. E. (1992). *Biology of the grapevine*. Davis, CA: Cambridge University Press.

Peynaud, E. (1983). *The taste of wine*. Paris: John Wiley and Sons.

Ray, E., & Marinacci, B. (2002). *Vineyards in the sky: The life of legendary Vintner Martin Ray*. Saratoga, CA: The Bookmill.

Smart, R., & Robinson, M. (1991). *Sunlight into wine: A handbook for winegrape canopy management*. Wauchope: Winetitles, Ministry of Agriculture and Fisheries, New Zealand.

Spayd, S. E., Tarara, J. M., Mee, D. L., & Ferguson, J. C. (2002). Separation of sunlight and temperature effects on the composition of Vitis vinifera cv. Merlot berries. *American Journal of Enology and Viticulture, 53*(3), 171.

Sullivan, C. L. (1982). *Like modern Edens: Winegrowing in Santa Clara Valley and Santa Cruz Mountains 1798-1981*. Cupertino, CA: California History Center.

Chapter 9
Quest for Quality Wine, Every Time: Guide to Root Cause Analysis. Summary and Conclusions

9.1 Summary

Together we have embarked on a quest to make quality wine. Our journey followed a path of organized and systematic steps. We recognized the road to making quality wine, every time, is truly complex. Our approach navigated the maze of possible problems. There are many variables and conditions that will create quality problems that every grape grower or winemaker will face at some point in their careers. This myriad of possibilities was parsed into six categories. Our investigations and systematic problem-solving methods were based upon proven rational approaches applied to the theory and practice of growing grapes and making wine. Table 9.1 is a summary of applied methods and some of their respective features and benefits.

Start by setting your quality goals. Understand the quality of your own wine and the quality of others. Include a broad array of quality indicators and wine characteristics in your own personal definition. Solicit and pay attention to independent reviews of wine critics. They can enhance your own. You can work to connect wine characteristics and wine quality proxies to help navigate your way during the grape growing and winemaking processes. Finding the root cause of a problem needs to start with a clear understanding of what the problem is and how you want it resolved.

We reviewed our basic winemaking practice to establish a common reference. A detailed problem statement should be established to launch an investigation into a wine quality problem. Document the various process phases.

Root cause analysis (RCA) is a set of methodologies to help identify all possible causes of a problem and then determine which one is the most likely root cause. The cause and effect diagram, also known as the fishbone diagram, lays out six major categories of causes which are processes, material, equipment, people, measurement, and environment. We have populated a fishbone with typical winemaking subcategories. We defined a very specific and clear problem statement which is intended to be the "effect" you are working to trace back to its root cause.

© The Editor(s) (if applicable) and The Author(s), under exclusive license to
Springer Nature Switzerland AG 2020
J. Steakley, B. Steakley, *A Quest for Quality Wine, Every Time*,
https://doi.org/10.1007/978-3-030-34000-1_9

Table 9.1 The features and benefits of the applied methods discussed in this book are summarized below. One or all of them make a powerful toolset to find the root cause of your problems and improve the quality of your wine

Method	Features	Benefit 1	Benefit 2
Define quality goals	Provide early clarity and specificity of the actual problem to solve	Minimizes wasted time and effort exploring irrelevant possibilities and details	Provides thoughtful personalized quality definition as reference to work toward and maintain focus
Cause and Effect Diagram and Analysis	Rational approach that initially spans all possible causes then narrows to the most probable cause	Clarity in problem statement provides specificity of effect to allow establishment of ties to probable causes	Fishbone visualization provides clear graphic of connections between cause and effect
RCA interrogation with the "5 Whys"	Penetrating questions elicit responses to better understand details of situation	Progressive questions build on response to dive deeper and reveal obscured layers of possible root cause	Progressive questions also build on response to focus in on possible root cause
Application of preventive measures	Use of preventive measures will minimize problems before they begin	Preventive measures will maintain and raise product quality	Preventive measures will establish high-quality standard for future vintages
Test your product	Chemical and sensory tests provide objective evidence	Sensory tests provide evidence in ways your product will be used	A broad spectrum of tests enable comparison with standards and insight into emerging trends
Kepner–Tregoe Analysis (KTA)	KTA is a rational approach that enables an efficient and value driven decision-making process	KTA matrix provides clear visualization and comparison of options and relative scoring	The KTA analysis approach can also incorporate possible adverse consequences and incorporate these into the decision-making process
Best of the best and worst of the worst (BowWow) analysis	BowWow uses key specific relevant history that focuses on your particular situation	BowWow effectively narrows the universe of possibilities to a smaller field of problem suspects	BowWow illuminates the contrast between the best aspects of the best product and the worst aspects of the worst product. Shines spotlight on probable root cause suspects

We discuss Kepner–Tregoe rational decision-making methods to aid in picking the optimal approach for resolving your winemaking problem. This technique allows for input of your own relative values and attributes you wish to be involved in the decision. You can weight these heavily toward quality and ignore cost and schedule, or you may have a budget or time budget and you still want some reasonable chance for improving quality. Given the uncertainties of the world we all live in, you can also include risk as a factor in making the decision.

When problems first show up and you need some help in getting started, you can employ BowWow (best of best and worst of worst) techniques. This tool will organize and sift through your information and identify the most likely problem causing suspects. You will have narrowed the field of root cause suspects and can focus your limited time and resources on probable suspects.

Use these tools to identify contributing factors and the root cause of your problem. Your winemaking problem-solving skills will improve. Someday soon, you will make your own quality wine. Objective improvements can mitigate marginal conditions and move your wine toward better balance and quality. Our quest has many layers. Solving winemaking problems is one. Making quality wine is another. Perhaps the most important is helping you feel the joy of creating something special.

9.2   Conclusion: Are We There Yet?

Wow, our 2016 Cabernet Sauvignon estate wine tastes great! The color, aroma, and taste are excellent. We grew the grapes, cared for the vines, managed the canopy, harvested the fruit, crushed the grapes, fermented the must, pressed, filtered, bottled, and stored the wine. We did it with a lot of help from friends and family. How could it not be good? We have made many mistakes but we have learned. We make fewer now. We pay close attention to the details. We think our 2016 estate Cabernet Sauvignon is a quality wine. We thought it was very good, but did other knowledgeable wine tasters and critics? How about independent wine experts? We entered our wine in a big well-known county fair home winemaker competition. This included blind tastings performed by expert judges.

Hooray! Our 2016 estate Cabernet Sauvignon did earn a gold medal. We smiled and laughed. We opened a bottle of our quality wine that night. It was delicious. It did indeed have the hue of a bright clean deep red Cabernet, it smelled of dark berries, tasted of dark cherry fruit, it was balanced, had good texture, and ended with a very nice quality finish. Not only was it delicious for a three-year-old Cabernet Sauvignon, we felt it was a bit astringent and we expect it to age well. To us, it was a true joy to drink. It felt like "sunlight held together by water" (Galilei, 1564–1642)[1].

When you have improved the quality of your wine, appreciate the accomplishment. Celebrate with friends and family. Recognize the achievement. Raise a glass, as in Fig. 9.1; to quality vineyards, grape growing, and winemaking. Enjoy what you have created. It is something very special.

---

[1] Galileo Galilei was a brilliant astronomer, physicist, and engineer. He made epic strides in observational astronomy and modern physics. He pioneered the experimental scientific method, confirmed Copernicus' theory that the earth revolved around the sun. He was a rebel. This same person provided us with a beautiful quote about wine: "*Wine is sunlight, held together by water*" Galileo was an inspiration.

Fig. 9.1 Raise a glass to quality vineyards, dedicated and rational grape growing, and creative and systematic winemaking. All three play a role in the making of a quality wine

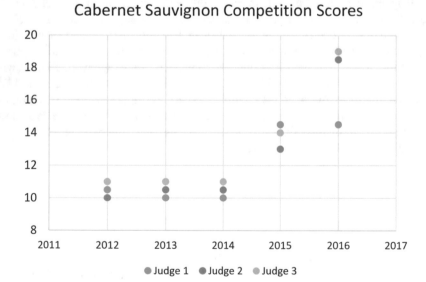

Fig. 9.2 Independent Wine Judges scores provide objective evidence of the Quality of your wine. They do not always agree; however, patterns may emerge, and trends will be revealed

9.2.1 Evidence of Quality Improvement

Use methods that are rational and systematic. They include really getting to know your own wine. It is so important to inspect, smell, and taste your wine. Look at it, smell it, and taste it. Grape growing and winemaking is very hard work, but not all the time. One very important piece of evidence of the quality of your wine is your own smile when you sample a barrel or open a bottle and pour a glass. You will know very quickly.

Additional useful feedback comes from scores, assessments, and descriptors from independent expert judges in blind tastings as shown in Fig. 9.2. Seek these out and participate often. The independence helps to eliminate bias and retain objectivity. The comments of expert critics provide rational and useful feedback. As we have mentioned, the score itself is only one aspect. It is the descriptive feedback in each of multiple categories that all are key to understanding the complexity and quality of your wine. Look for patterns, trends, and useful aids in improving next year's vintage. Not only is there joy in sharing a quality wine with friends, there is happiness and pleasure in the quest itself. You are well on your way to making quality wine, every time.

Our wine has improved, we are happy, and yet our quest for quality wine rages on toward new heights.

Sincerely, Joyce and Bruce.

Printed in the United States
by Baker & Taylor Publisher Services